Mechanisms and Machine Science

Volume 87

This book series establishes a well-defined forum for monographs, edited Books, and proceedings on mechanical engineering with particular emphasis on MMS (Mechanism and Machine Science). The final goal is the publication of research that shows the development of mechanical engineering and particularly MMS in all technical aspects, even in very recent assessments. Published works share an approach by which technical details and formulation are discussed, and discuss modern formalisms with the aim to circulate research and technical achievements for use in professional, research, academic, and teaching activities.

This technical approach is an essential characteristic of the series. By discussing technical details and formulations in terms of modern formalisms, the possibility is created not only to show technical developments but also to explain achievements for technical teaching and research activity today and for the future.

The book series is intended to collect technical views on developments of the broad field of MMS in a unique frame that can be seen in its totality as an Encyclopaedia of MMS but with the additional purpose of archiving and teaching MMS achievements. Therefore, the book series will be of use not only for researchers and teachers in Mechanical Engineering but also for professionals and students for their formation and future work.

The series is promoted under the auspices of International Federation for the Promotion of Mechanism and Machine Science (IFToMM).

Prospective authors and editors can contact Mr. Pierpaolo Riva (publishing editor, Springer) at: pierpaolo.riva@springer.com

Indexed by SCOPUS and Google Scholar.

More information about this series at http://www.springer.com/series/8779

Andrii Kudriashov · Tomasz Buratowski ·
Mariusz Giergiel · Piotr Małka

SLAM Techniques Application for Mobile Robot in Rough Terrain

Springer

Andrii Kudriashov
Department of Robotics and Mechatronics
AGH University of Science
and Technology
Kraków, Poland

Tomasz Buratowski
Department of Robotics and Mechatronics
AGH University of Science
and Technology
Kraków, Poland

Mariusz Giergiel
Department of Robotics and Mechatronics
AGH University of Science
and Technology
Kraków, Poland

Piotr Małka
Department of Robotics and Mechatronics
AGH University of Science
and Technology
Kraków, Poland

ISSN 2211-0984 ISSN 2211-0992 (electronic)
Mechanisms and Machine Science
ISBN 978-3-030-48980-9 ISBN 978-3-030-48981-6 (eBook)
https://doi.org/10.1007/978-3-030-48981-6

This Springer imprint is published by the registered company Springer Nature Switzerland AG
The registered company address is: Gewerbestrasse 11, 6330 Cham, Switzerland

Preface

The development of SLAM-based mobile robot control system as the integrated approach, which connects localization, mapping and motion control fields, is presented in this book. The author has performed a survey and selection of several techniques that represent the basics of the mathematical description of wheeled robots, their navigation and path planning approaches, localization and map creating techniques. SLAM paradigms and Bayesian recursive state and map estimation techniques that include Kalman, and particle filtering were studied in a detailed way in a separate chapter of this book. These fundamentals allowed having the development of SLAM-based integrated system for the inspection task performed. The system development proposed by the author is divided into two phases: a single robot approach and multirobot inspection system. The first phase has been performed in three steps: the development of pose tracking technique with a focus on rough terrain application; the implementation of combined 2D and 3D mapping application that might be used in outdoor, indoor and multi-level terrain; and the integration previous steps implementation into one solid system. In pose tracking phase, an original hybrid approach that connects AMCL and EKF filtering paradigms for the local state estimation in the local frame has been presented. Mapping is separated into 2D, which is performed by the Rao-Blackwellized 2D SLAM technique, while for the 3D workspace recreation, the algorithms that deliver required for 3D mapping pose and measurement were developed. The original approach for 2D SLAM in multi-floor building that covers each 2D level map, and continuous 3D pose tracking can be found in this book. The multirobot inspection system after study of multiagent system and robots cooperation is considered a group of homogeneous mobile robots. The last part of research is dedicated to multirobot map creation and path planning solution development. Since occupancy grid map can be recognized as an image, the merge of maps can be considered an image processing task. The author applied decentralized map fusion technique which is based on the recent pose and map exchange during sporadic robot meetings—the "rendezvous events". The robots' homogeneous behavior and configuration allow to develop a multirobot system without theoretical limitations of the number of used robots, which may be added to a group at any time.

The evaluation of the integrated SLAM-based mobile robotic solution developed in this book for inspection task performance in single or multiagents approaches has been performed in two phases. The first one is a simulation verification by V-REP simulator usage, and the second an experimental verification, prepared in the rough underground terrain of an experimental mine located at the Faculty of Mechanical Engineering and Robotics of the AGH University of Science and Technology in Krakow. The experiment results confirm the robustness and efficiency of the developed SLAM-based integrated mobile robot approach for inspection tasks in a rough terrain.

Kraków, Poland

Andrii Kudriashov
Tomasz Buratowski
Mariusz Giergiel
Piotr Małka

Contents

Chapter 1
Introduction

Abstract Chapter 1 presents the book research motivation, which contains a brief explanation of the state of modern mobile robotics. Then, the book Statement has been formulated. The last section presents a defined aim and scope along with the book organization structure.

Keywords Mobile robotics · SLAM · Mapping · Robot localization · Inspection robots · Authonomous robots

1.1 Motivation

Many things have changed since the time when, in the early 1920s, Karel Čapek first called a mechanical creature a 'robot'. The word 'robot' originates from a Czech word, which means 'work'. These machines, which are "mechanically perfect" and have "an enormously developed intelligence", as one of the main heroes in Čapek's R.U.R. describes them [2], in a very short time, became an inspiration and one of the main research fields in mechanical and artificial intelligence science. During many decades, they consequently went from science-fiction protagonists or antagonists, through the first laboratory prototypes, to efficient 'workers' in probably all engineering fields, especially industrial ones. Novel technologies and a rapid scientific progress resulted in the big robotics 'revolution' at the turn of XX and XXI centuries, which introduced robots to the social life of many countries. One of the most rapidly developing fields in robotics science is mobile robots, which are constructed to perform different tasks, such as inspection or servicing. Currently, modern mobile robots can perform the tasks of unknown space explorers, ordinary home cleaners, medical servants in hospitals or even pets for children. And they are no longer as exotic or expensive as they were before.

This book discusses mobile robotic approaches with the focus on inspection robots for work in terrain, which may be hard-to-access or even dangerous for human life or health. One of most important characteristics of this type of robotic systems is their ability to work autonomously. This means that a mobile robot should be able to perform tasks in unknown terrain without continuous human supervision

A. Kudriashov et al., *SLAM Techniques Application for Mobile Robot in Rough Terrain*, Mechanisms and Machine Science 87, https://doi.org/10.1007/978-3-030-48981-6_1

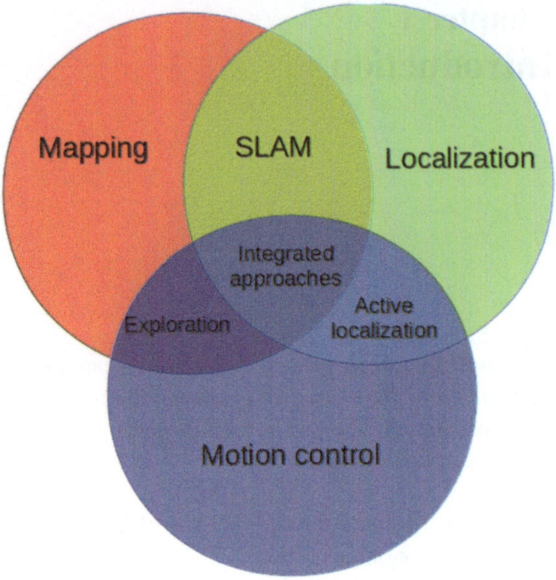

Fig. 1.1 The fields of mobile robotics connected by the integrated approach

or remote control. This requires for many different problems like mapping, precise localization and motion control to be solved. According to Makarenko et al. [1], the mobile robotics fields, which cover the mentioned problematics, can be connected by separate techniques, such as simultaneous localization and mapping (SLAM), exploration and active localization. The integration of these three fields gives a group of different robotic techniques with a common name of the integrated approaches (Fig. 1.1).

Since each of the mentioned fields consists of a huge number of different techniques that may look very efficient in one case but weak in others, their integration requires a deep study of the fundamentals and details of the selected techniques along with the classification of the recommenced workspace features. Therefore, in many cases, the discussion on which selected technique composition and integration approach will be the best or 'most efficient' seems to be more philosophical than engineering-related. There is no ideal approach that can satisfy all or most of the possible requirements, so the selection must be considered as a set of compromises applied to a specific workspace and the performed task.

The integrated approach development usually starts from a survey on the techniques that represent a separate part of the approach. For autonomous inspection systems, SLAM seems to be the most important, as it is a product of localization and mapping approaches, as well as navigation, which includes path planning in chosen world representation and the basics of motion control. The classic and the easiest localization techniques are based on incremental methods that are represented by odometry and inertial navigation systems. Odometry is usually based on measurements from deployed in wheel motor encoders, so it gives fast and low-cost information of the path passed by the robot, but can be very imprecise in motion on surfaces where skid and slippage effects are present. In light of this, different odometry methods that exclude wheel-to-surface contact were developed in recent years. The most known of the mentioned techniques are visual odometry (VO) and laser odometry (LO), which find their practical application even in space exploration programs. The integrated approach based on visual odometry could be found in NASA's Mars rover mission data report published by the name, "Two Years of Visual Odometry on the Mars Exploration Rovers" [3]. Laser odometry has similar basics, but unlike VO, it assumes the use of laser ranger-finder scanners instead of visual camera sensors. This may allow for faster reactions and more precise observations in low-light workspaces, but the measurements are monochrome and may be applied only in cases where the environment representation only requires precise shapes. Since each of these methods has different strong and weak sides, the multidomain data fusion and localization estimation is considered as the regular approach in mobile robotics practice. The two main approaches to localization estimation, which can be found in robotics literature, are the Gaussian Kalman filter and the particle filter estimator. Kalman allows to fuse data from many sensors, but based on all previous estimates and gathered measurements, it can be considered as weak point in some cases, since huge measurement failures will have an impact on all subsequent estimates, and the general accuracy of filtering will be decreased. The other approach—particle filter—does not experience this issue, but it allows data inputs from only one measurement source. Therefore, both filtering techniques found high use in mobile robotics.

Mapping approaches are generally divided into a planar and a 3D representation of the world. Each type can be used for both navigation and workspace state recreation purposes. The efficiency of motion planning strictly depends on the map type and accuracy, so they must be defined by a chosen navigation approach. The motion planning technique, in turn, depends on the kinematic constraints of the mobile robotic platform. Thus, the planar representation could be more useful for wheeled robots, while unmanned aerial vehicles (AUV) require three dimensions. Because this book is focused on wheeled robots, the planar world representation seems to be more helpful. From the inspection point of view, unlike the one mentioned previously, the general picture of the inspected environment represented by three dimensional measurements is definitely more informative and finds wider usage. Thus, the hybrid solution might be an interesting compromise for inspection and exploration tasks.

The exploration time, as well as the time of performing a task, can be significantly decreased through the application of a group of mobile robots. Cooperation in a group is a separate field of robotic science and requires an additional study on the nature of the group behavior. Since the main focus in this book is on autonomous work performed by a mobile robot in unstructured rough terrain with the use of a low number of sensors, the possibility of a robot cooperation in a group is an additional task.

Therefore, from the thus far presented brief description of the mobile robotics field, the book Statement can be formulated.

1.2 Book Statement

Based on the mobile robotics brief presented in the motivation, the following definition of the book statement can be proposed:

'It is possible to develop an effective mobile robots inspection system, which is based on simultaneous localization and mapping techniques (SLAM) with combined Adaptive Monte-Carlo localization and Extended Kalman filter (AMCL-EKF) laser odometry pose estimation along with 2D and 3D mapping, and which may be applied for a group of mobile robots for a rough terrain tasks'.

1.3 Aim and Scope

The aim of this book is present the development of a SLAM-based mobile robot control integrated approach system. Based on the Motivation, the research performed in this book includes the following steps: a survey on the existing techniques of motion control; localization and mapping with a focus on the application of integrated solution; a study of SLAM fundamentals and selected techniques; a selection of the most applicable techniques in accordance with the book Statement and a study of their integration into one of the solution possibilities; the development of a SLAM-based mobile robot control system for inspection tasks in unstructured rough terrain; a study of multirobot approaches and the development of system for a single robot integration into a group of mobile robot inspection system; and an evaluation of the proposed methods and systems.

Therefore, the book research can be covered by the following chapters:

Chapter 1 presents the book research motivation, which contains a brief explanation of the state of modern mobile robotics. Then, the book Statement has been formulated. The last section presents a defined aim and scope along with the book organization structure.

Chapter 2 can be considered as an introduction to mobile robotics. This chapter covers a brief mathematical description of mobile robots that consists of kinematic and dynamics with non-holonomic constrains applied to wheeled robots. Then, a terrain representation and mapping survey has been conducted. After that, the path planning study was performed. It includes a planner definition and several approaches to navigation, which are divided by different behaviors and types of world representation. The last part is dedicated to localization approaches and covers the local incremental method and global localization techniques.

In **Chap.** 3, SLAM is considered as a probabilistic approach that originates from Bayes rule and Markov assumption. The estimation techniques for the robot's pose and map are presented as parts of a probabilistic framework. The Bayesian recursive estimation definition and the Bayes filter implementations, like Kalman and discrete particle filters, were studied here. The map estimation is based on three- and two-dimensional occupancy grids. The final part of this chapter connects these two estimation values into a common one technique that performs they simultaneously. Three different SLAM paradigms—EKF-based, particle-based and Graph-based or Topographic SLAM—has been discussed.

Chapter 4 presents the development of an original 2D/3D SLAM application solution for inspection tasks. This chapter covers a group of mobile robot fundamentals and defines the chosen paradigm of robots in group cooperation behavior. Then, the integrated approach development for a single robot is presented. The SLAM-based control system is developed in accordance with the book Statement and includes a combined 2D and 3D mapping approach with the original robust pose estimation technique proposition that connects the EKF and AMCL techniques. The development covers tasks in structured and unstructured rough terrain and the original multilevel building inspection approach. The final part presents the developed system application for the multirobot inspection technique. Different architectures were discussed and the implementation of a multi-agent inspection system proposition has been developed.

In **Chap.** 5, the developed methods as well as an integrated single and multirobot systems evaluation can be found. This chapter is divided into two parts a simulation verification and an experimental evaluation. In the first part, several simulation scenes were created in the V-REP environment. This verification covers all parts of the proposed system for both single and multirobot cases. The experimental part has been performed in the form of an inspection and rescue operation in real rough underground terrain, which can be found in coal mines after accidents. For this purpose, the off-road mobile robot platform and a 3D laser rangefinder scanner have been used.

Chapter 6 closes the book research by summarizing the results and discussing other conclusions. A possible future work direction and research plans can be found in chapter.

References

1. Makarenko, A.A. et al.: An experiment in integrated exploration. In: 2002 IEEE/RSJ International Conference on Proceedings of the Intelligent Robots and Systems, vol. 1, pp. 534–539. IEEE (2002)
2. Čapek, K.: RUR:(Rossum's Universal Robots): a Play in Three Acts and an Epilogue. Humphrey Milford, Oxford, UK (1928)
3. Maimone, M., Cheng, Y., Matthies, L.: Two years of visual odometry on the mars exploration rovers. J. Field Robot. **24**(3), 169–186 (2007)

Chapter 2
Introduction to Mobile Robots Navigation, Localization and Mapping

Abstract This chapter can be considered as an introduction to mobile robotics. This chapter covers a brief mathematical description of mobile robots that consists of kinematic and dynamics with nonholonomic constrains applied to wheeled robots. Then, a terrain representation and mapping survey has been conducted. After that, the path planning study was performed. It includes a planner definition and several approaches to navigation, which are divided by different behaviors and types of world representation. The last part is dedicated to localization approaches and covers the local incremental method and global localization techniques.

Keywords Kinematics and dynamics · Navigation · Localization · Mapping · Mobile robots fundamentals

2.1 Introduction

This chapter provides an overview on mathematical description and several mobile robotics methods selected from fields, which are covered by integration approaches, as important background for control applications based on SLAM techniques. The review begins with mathematical description basics that are required for robots control and modeling. It is done by walk through kinematics constrains presented in wheeled mobile robots constructions and the kinematic equations formulation along with the review of several dynamics description formalisms for the differential drive mobile robots. Then the survey on robots navigation methods that covers path planning fundamentals and different techniques discussion including their dependency on the presented different types of world representations can be found in this chapter. The final part is dedicated to localization techniques involving incremental methods and global positioning approaches.

A. Kudriashov et al., *SLAM Techniques Application for Mobile Robot in Rough Terrain*, Mechanisms and Machine Science 87, https://doi.org/10.1007/978-3-030-48981-6_2

2.2 Mathematical Description of Mobile Robots

For an efficient control of mobile robot the correct mathematical model that describes its mechanics is required. Mobile robots can use very wide types of propulsion from simple wheels to nature inspired leg-like pedipulators of large complexity. This book is focused on wheel-based, differential drive solutions. Thus, in this chapter mobile robots that use wheels are assumed.

2.2.1 *Holonomic and Non-holonomic Mobile Robots*

One of the most important characteristics of mobile robots that is a number of possible degrees of freedom (DOF) which can be operated directly ('in a place') compared to the whole number of DOF. If these values are equal, that type of dynamic systems is called *holononomic*, and if they don't the systems are *non-holonomic*. For example in two-dimensional space human can be described as holonomic, but when we place a man in usually observed three dimensions, he becomes a non-holonomic, as people cannot fly. Moreover, holonomic constrains could be removed by the reduction of number of dimensions, but non-holonomic constrains cannot.

This property could be described in more accurate way by the example of single-wheeled robot in two-dimensional space [1, 2].

If a robot is described by x, y position and θ orientation (Fig. 2.1), then its motion constrains can be described as following:

$$\dot{x}\sin\theta - \dot{y}\cos\theta(+d\theta \cdot 0) = 0 \tag{2.1}$$

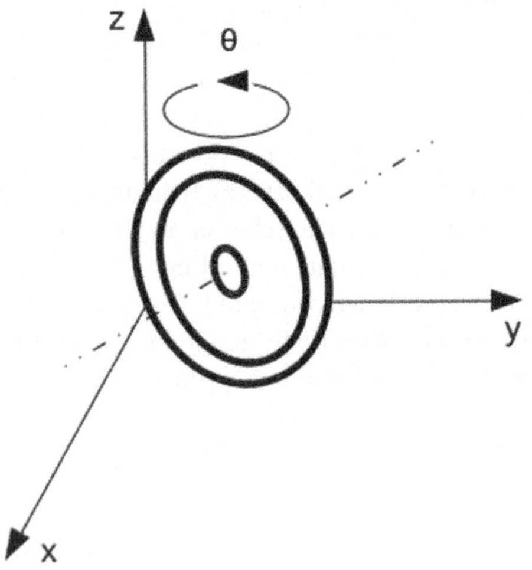

Fig. 2.1 The single-wheeled mobile robot description in xy plane

An Eq. (2.1) is related to Pfaff's transformation:

$$\mathbf{A}(q)\dot{q} = 0 \tag{2.2}$$

where **A** is a full-rank matrix, that in accordance with (2.1) looks as follows:

$$\mathbf{A}\left(\begin{bmatrix} x \\ y \\ \theta \end{bmatrix}\right) = \begin{bmatrix} \sin\theta \\ -\cos\theta \\ 0 \end{bmatrix} \tag{2.3}$$

The system is holonomic if for (2.2) it is possible to find the vector of function $\Phi(X)$ that makes condition (2.4) true, else it is non-holonomic:

$$\frac{\partial\Phi(q)}{\partial q} = \mathbf{M}(q)\mathbf{A}(X) \tag{2.4}$$

which for a single-wheel robot, if the function exists, it has the following form:

$$\frac{\partial\Phi(q)}{\partial q} = \begin{bmatrix} \frac{\partial\Phi}{\partial x} \\ \frac{\partial\Phi}{\partial y} \\ \frac{\partial\Phi}{\partial\theta} \end{bmatrix} = m(q)\begin{bmatrix} \sin\theta \\ -\cos\theta \\ 0, \end{bmatrix} \tag{2.5}$$

where $m(X) \neq 0$. The calculation of first two conditions from (2.5) gives false result, thus, it might be concluded that the presented construction is non-holonomic. However, if we assume that $\theta = \theta_0 = const$, which means that the turning is limited to fixed radius, then the motion condition (2.1) will look as follows:

$$\dot{x}\sin\theta_0 - \dot{y}\cos\theta_0 = 0 \tag{2.6}$$

This could be interpreted as robot can follow only a straight and for $\theta_0 = 0$ system will be holonomic, since the function that makes condition (2.5) true exists. The presented example confirms that holonomic constrains can be removed by dimensions reducing.

Let's look at the simplest two-wheeled construction placed at xy plane (Fig. 2.3).

It is very easy to notice that because of the construction properties of the used wheels, the presented model is non-holonomic. Although robot's state space has DOF number 3, the robot is definitely limited to move only on straight across x-axis and its rotation can be reached by differences in each wheel velocities. That kind of drive is called as *differencial* and its kinematic description is presented in the next section of this chapter.

The holonomics directly impacts on controls and path planning. For non-holonomic systems it is more difficult and it takes more time to find proper, shorter and efficient trajectory that avoids discovered obstacles. Let us imagine a four wheeled vehicle with fully controlled 2 DOF—forward/back and forward wheels rotation placed in two-dimensional space with a set of obstacles shown on Fig. 2.2. The trajectory from

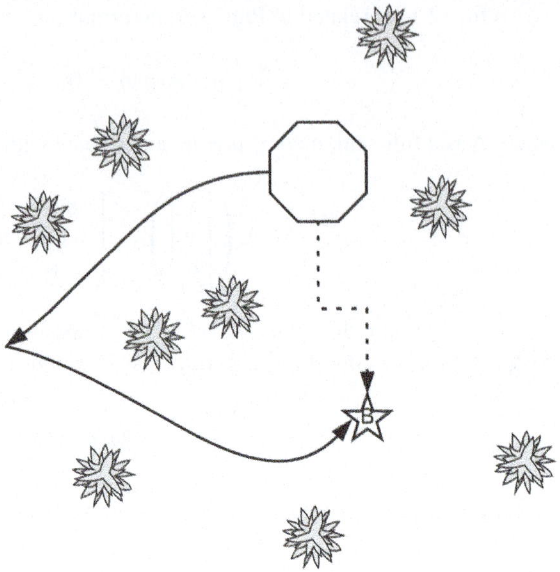

Fig. 2.2 Non-holonomic robot control: shaped path is allowed for holonomic and solid for non-holonomic robots

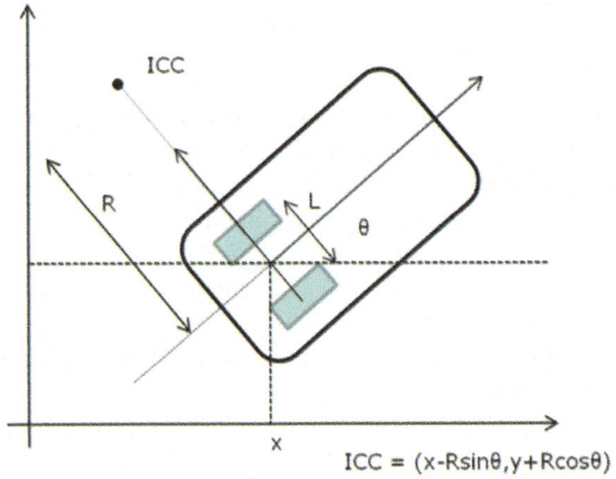

Fig. 2.3 Two-wheeled mobile robot model description in two-dimensional space

current location to point B is not short as it could be in holonomic construction case, when the vehicle would be able to slide left or right.

The next part of this section is focused on mathematical description of differential drive wheeled mobile robots.

2.2.2 Kinematics

The first important description of any dynamic systems is a kinematic model [1, 3, 4]. Here, two kinematics problems to solve can be defined—*forward*, that describes position and orientation of each characteristic point of system with respect to reference frame and *reverse*, that reconstructs control parameters (angles, velocities and accelerations) from given trajectory.

The description of kinematics of any robot begins with defining characteristic points and their connection to a coordinate frame. The differential drive mobile robots kinematic model consists of the characteristic points: robot's frame central point, right and left wheel centers of the symmetry and Instantaneous Center of Curvature (ICC)—a point that lies along common wheel axis and about which robot rotates; the geometric parameters: the distance between wheel centers (l), wheel radius ($r = r_{right\ wheel} = r_{left\ wheel}$); the kinematic parameters: linear and angular velocities of each wheels (V_r, V_l and ω_r, ω_l), the distance between points A and ICC (R), the angle that describe robot's frame orientation relates to the world frame (θ) and its rotation speed ($\dot{\theta}$); coordinate frames: world (x_0, y_0, z_0) and robot (x, y, z). As robot rotates about ICC and this rotation ω is the same for the whole construction, we can describe each wheel velocities relation as follows:

$$\begin{aligned} V_r &= \omega(R + \tfrac{l}{2}) \\ V_l &= \omega_(R - \tfrac{l}{2}) \end{aligned} \tag{2.7}$$

Thus, the rotation speed might be described as follows:

$$\omega = \frac{V_r - V_l}{l} \tag{2.8}$$

with distance to ICC as:

$$R = \frac{1}{2} l \frac{V_l + V_r}{V_r - V_l} \tag{2.9}$$

Assuming that robot is in a stated pose related to a coordinate frame the location of ICC will be:

$$ICC = (x - R\sin\theta, y + R\cos\theta) \tag{2.10}$$

with (2.8) and (2.9) it is easy to describe robot's linear velocity as well:

$$V = \omega R = \frac{1}{2}(V_r + V_l) \tag{2.11}$$

Finally, the *forward* kinematics of differential drive in a world frame for the time t with given linear (2.11) and angular (2.8) velocities can be described as follows:

$$\begin{bmatrix} \dot{x_0} \\ \dot{y_0} \\ \dot{\theta} \end{bmatrix} = \begin{bmatrix} \cos\theta(t) & 0 \\ \sin\theta(t) & 0 \\ 0 & 1 \end{bmatrix} \begin{bmatrix} V(t) \\ \omega(t) \end{bmatrix} \tag{2.12}$$

From a mobile robot control perspective, the relation between robot's base linear and angular to angular velocity of each wheel is more valuable. For that purpose knowledge about wheel parameters is required, and the kinematic model of differential drive in robot's frame is described as follows:

$$\begin{bmatrix} V_x \\ V_y \\ \dot{\theta} \end{bmatrix} = \begin{bmatrix} \frac{r}{2} & \frac{r}{2} \\ 0 & 0 \\ -\frac{r}{2}l_1 & \frac{r}{2}l_1 \end{bmatrix} \begin{bmatrix} \omega_l \\ \omega_r \end{bmatrix} \tag{2.13}$$

Equation (2.13) is very important and useful for creating an advanced control systems.

2.2.3 Dynamics

Since the kinematic model does not include the forces that affect motion, they are considered in dynamics modeling. Lagrangian and Newtonian approaches are usually used for motion equations formulation. Since this book does not focus on low-level motion controllers and dynamics models are usually excluded from path planning approaches by assuming that drive controllers used in mobile robots are constructed well enough and can perform safe acting with avoiding dynamics disasters (Sect. 2.4), this part of book presents only general principles of wheeled robots dynamics without detailed formulations.

Lagrangian method formulates dynamic model in a following way:

$$\frac{d}{dt}\left(\frac{\partial \mathcal{L}}{\partial \dot{q}}\right) - \frac{\partial \mathcal{L}}{\partial q} = \mathbf{Q} - \mathbf{A}^T \lambda \tag{2.14}$$

where $\mathcal{L} = E - V$ is a Lagrange function which equals to difference between kinetic E and potential V energies of system, A is Jacobian matrix and λ is a vector of Lagrange multipliers. Generalized coordinates of system are described as vector

$$q = \left[x \; y \; \theta \; \phi_l \; \phi_r\right]^T \tag{2.15}$$

where Q is a vector of generalized forces that affect dynamic system. The Lagrangian description begins with left side of Eq. (2.14) by solving for each coordinates in q. This way of description is one of most popular in mechanical science literature [5, 6].

Another method of description, based on Lagrangian equations, is called Maggi dynamic equations [5–7]:

$$\sum_{j=1}^{n} C_{ij} \left[\frac{d}{dt} \left(\frac{\partial E}{\partial \dot{q}_j} \right) - \left(\frac{\partial E}{\partial q_j} \right) \right] = \Theta_i \tag{2.16}$$

$$\dot{q}_j = \sum_{i=1}^{s} C_{ij} \dot{e}_i + G_j \tag{2.17}$$

where i is related to number s of independent parameters in $\dot{q}$ that equals to number of system's degrees of freedom, j to number n of generalized coordinates, C_{ij}—coefficient matrix, G_j—coefficients, Θ_i—generalized forces, e_i—characteristic of kinematics of the system in generalized coordinates. By applying definition of virtual work of parameters $\delta \dot{e}_i$ to (2.16) the Maggi's formulation could be extended to external forces Q_j as

$$\sum_{i=1}^{s} \Theta_i \delta e_i = \sum_{i=1}^{s} \delta e_i \sum_{j=1}^{n} C_{ij} Q_j \tag{2.18}$$

which in matrix form can be written as [1]

$$\sum_{j=1}^{n} C_{ij} L_j = \Theta_i \;{}_{i=1..s} \tag{2.19}$$

where L is related to matrix form of dynamic equation and defined as

$$L = M(q)\ddot{q} + C(q, \dot{q})\dot{q} \tag{2.20}$$

Coefficients $\dot{e}_i$ and G_j for two-wheeled robots equals [5]

$$\begin{aligned} \dot{e}_i &= \left[\dot{\phi}_l \; \dot{\phi}_r \right] \\ G_j &= \left[0\,0\,0\,0\,0 \right] \end{aligned} \tag{2.21}$$

As dynamics of the system described by Maggi's formalism are not affected by Lagrange multipliers and number of equations equals to number of system's DOF the formulation of dynamics description of the system can be easier in many cases. Therefore, it is widely use in many mobile robotics works. There are also many different formulations of dynamic systems like Appell's equations [8, 9], Kane's method [10, 11] or multibody dynamics method [12, 13]. However, if moments of force in wheeled robot motors are most valuable, the Maggi's equations seem to be more convenient formulation of dynamics of the mobile robot [7].

2.3 Terrain Mapping

Terrain mapping is a task of workspace recreation by chosen, most efficient for the case, world representation model. In accordance to [14] types of world representation may be divided into two different approaches—topological and metric. However, in practice, a hybrid approach that connects those two main maps is often applied [15–18].

2.3.1 *Topological Maps*

Topological maps are represented as locations connected to one another. Probably one of the most frequently seen example of topological world representation is a city bus routes map (Fig. 2.4). It can be considered as a graph, where vertices or nodes are associated to exact places (stops in Fig. 2.4) while vertex edges are related to possible physical routes between these locations.

Topological maps are closely related to metrical, as during graph construction real-world metric measurements are used for delivering possible transition between node vertices (Fig. 2.5) [20].

Voronoi diagrams is another approach associated with world representation by topology [21]. This method is used for map building with minimal safe distances

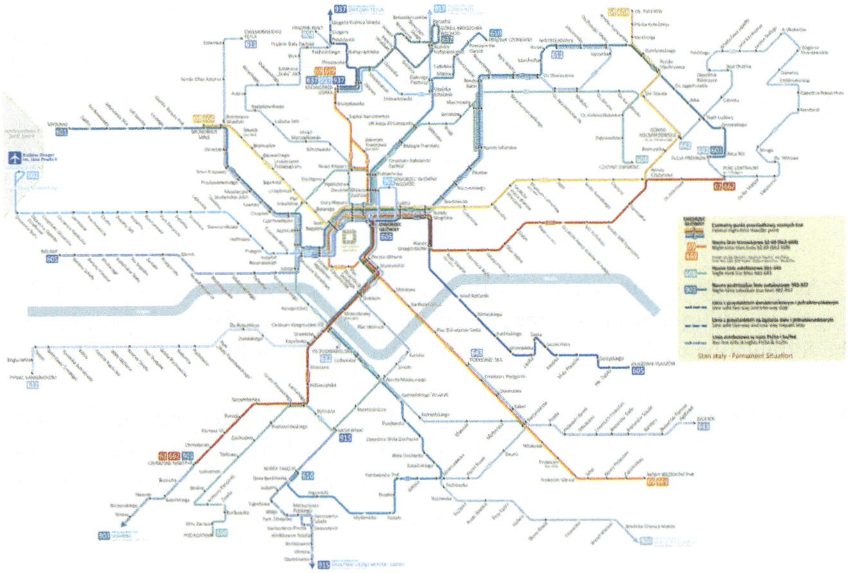

Fig. 2.4 Topological map example: night bus routes in the city of Krakow [19]

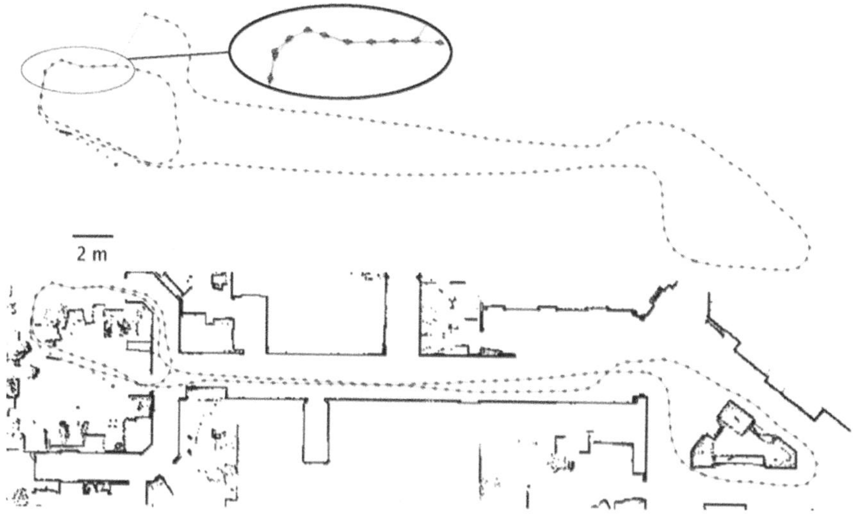

Fig. 2.5 Topological map constructed during indoor exploration (top) and its relation to metric (bottom) [20]

to the obstacles and is often used for mobile robot's navigation roadmaps. Voronoi diagrams as roadmaps are presented in Sect. 2.4.3.

2.3.2 Metric Maps

According to the way of representation, all metric maps can be divide into two main types—a set of *features* or *locations* [14, 22, 23].

Location-based map is described as a list of N elements related to each coordinates x, y and z including occupied as well as free spaces:

$$m = \left\{ m_{x_1,y_1,z_1}, m_{x_2,y_2,z_2}, \cdots, m_{x_N,y_N,z_N}, \right\} \tag{2.22}$$

The most widely used metric location-based map family is *occupancy* or *binary grid* maps. First binary grids model mapping algorithms was presented in late 80s by Moravec [24] and Elfes [25]. The authors used wide-angle sonar as sensing source, but nowadays the laser scanning is a dominating technique in binary grids mapping approaches. The idea is to divide observed space into binary cells forming rigid grid (Fig. 2.6).

The accuracy in this approach is directly depended on grid size and is a subject of compromises. The small cells give more accurate world representation with tiny details but cost a lot of computing loading and often are just not worthwhile in large empty spaces. On the other hand, for huge free space big cells seems to be more effi-

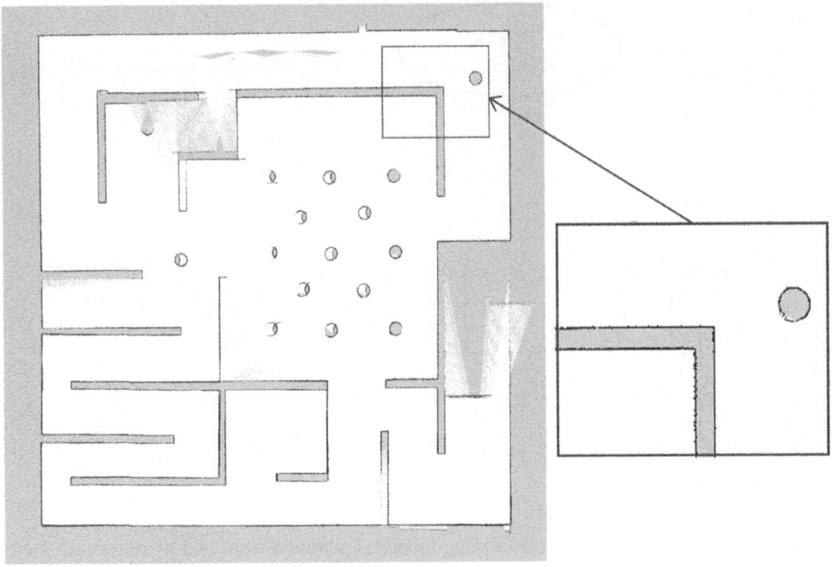

Fig. 2.6 An example of map based on occupancy grids

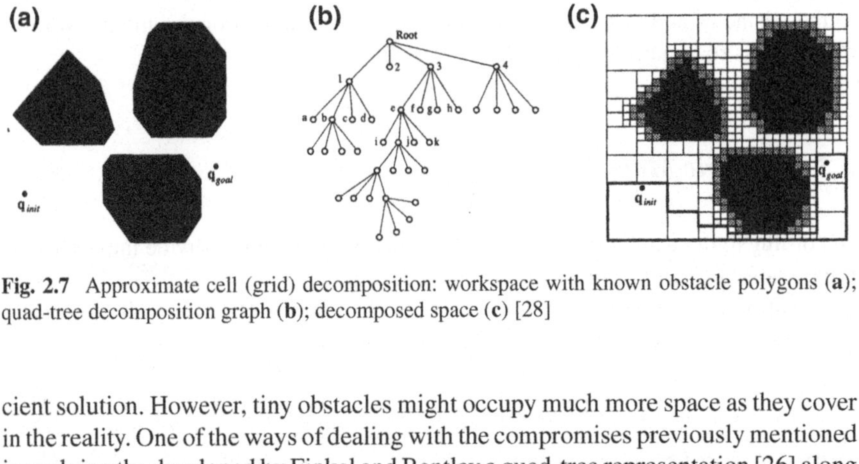

Fig. 2.7 Approximate cell (grid) decomposition: workspace with known obstacle polygons (**a**); quad-tree decomposition graph (**b**); decomposed space (**c**) [28]

cient solution. However, tiny obstacles might occupy much more space as they cover in the reality. One of the ways of dealing with the compromises previously mentioned is applying the developed by Finkel and Bentley a quad-tree representation [26] along with its later case-related optimizations [27]. The workspace and its representation by a *constant grid* with quadtree decomposition is shown on Fig. 2.7. The map might be already known or discovered/updated during goal-following motion with marking new obstacle cells black.

Another type of metric map representation is a list of landmarks, called *features*:

$$m = \{l_1, l_1, \cdots, l_N\}^T \tag{2.23}$$

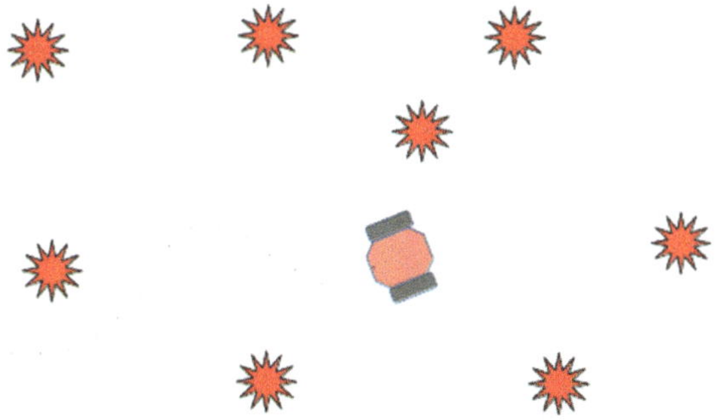

Fig. 2.8 Feature based map landmarking example—*trees* and marked robot's path (red) [29]

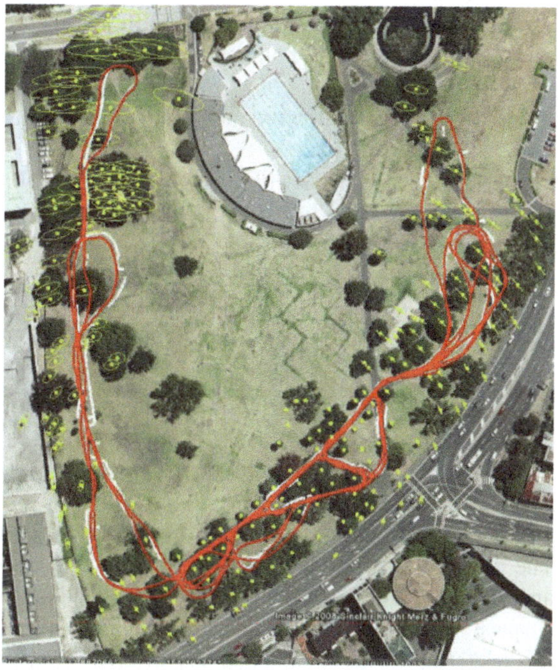

Fig. 2.9 An example of feature based map with extracted landmarks—*trees* and marked robot's path (red) [22]

Unlike the occupancy grids, which are dense and store information about free as well as occupied locations, the feature based maps rare and store only already discovered obstacles as landmarks with updates performed during exploration by appending newly discovered landmarks into the list that represents a map (2.23) (Figs. 2.8 and 2.9).

For feature based maps the landmark estimation by Kalman and informative filters is natural and robust. Their fundamentals are discussed in Chap. 3.

The presented above types of world representation are necessary for an efficient navigation of mobile robots and define the type of possible path plannning approaches that could be applied in the target workspaces. The discussion on this field of mobile robotics is presented in the next part of this chapter.

2.4 Mobile Robot's Navigation

2.4.1 Path Planning

A study of robots motion planning began in times when first robots in their modern definition did not start their autonomous moving yet. The robotics and AI communities developed a rich number of different theoretical and practical navigation approaches [31, 32]. The path planning definition might be formulated in a sentence as follows: "*a finding safe obstacle-free road from initial state to target*". This task solution depend on both word and robot configurations and their way of representations.

World, in accordance with the type of terrain, might be divided into two large groups—structured and unstructured. The first is defined mainly by man-created structures like roads or buildings where navigation problem is limited to finding and following this possible roads. The unstructured word is more complicated, since planning problem is multiple and has different origins. The two origins will be described and the attempt to solve them is one of the aims covered in this book. First origin, which is related to a way of world representation is briefly discussed in the previous Sect. 2.3. A survey on dedicated approaches of continuous path finding could be found in the current Section. The second origin is associated with a state estimation or localization problem, which is much more complex in a rough terrain. Chapters 3 and 4 are focused on this problem.

Robot and word configurations are connected by *Configuration Space* concept. The definition of robot's Configuration Space $\mathcal{C}$, also called $\mathcal{C}$*-space* is critical for any efficient path planning algorithms independent from robot's DOF number. The $\mathcal{C}$-space is a n-dimensional space that includes all robot $\mathcal{A}$ possible configurations in a operating world $\mathcal{W}$, and that has different from classical Cartesian, where n is a robot's DOF number [33–35]. In a motion planning C-space consists of free space and obstacle spaces $\mathcal{C} = \mathcal{C}_{free} \bigcup \mathcal{C}_{obs}$. Therefore, the path planning task might be defined as finding a continuous trajectory in $\mathcal{C}_{free}$ (Fig. 2.10). For two-dimensional space robot's third coordinate—orientation is usually skipped due to the huge computation requirements. This causes complications with robot's footprint construction since it must be radially symmetric and requires approximation.

Another computation costs reducing approach that is very often applied regards to dynamic constrains which might be skipped with the assuming that robot's controller is able to deliver acting on a level which guarantees safe robot's motion. This assumption is applied to this book, and dynamic constraints are not included into path planning approaches.

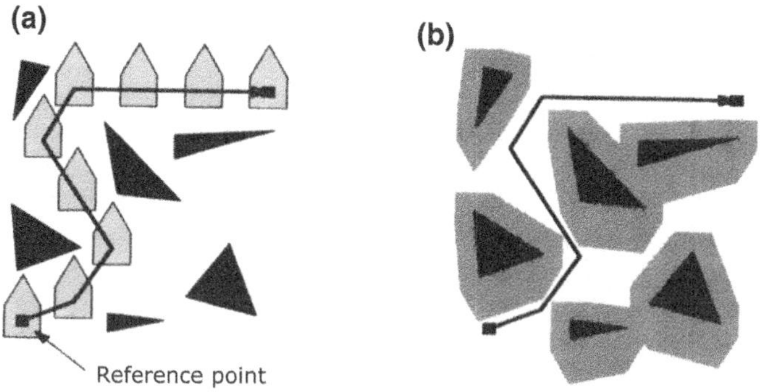

Fig. 2.10 Robot work space with obstacles (**a**) and its configuration space representation (**b**) [35]

From just learned basics it is easy to find that all approaches depend on world representation types—metric, topological or hybrid. However, in most cases (since grid is also actually a graph with fixed topology) the possible path in $\mathcal{C}$ can been considered as a graph

$$\mathcal{G} = (\mathcal{V}, \mathcal{E}, W_{\mathcal{E}}) \tag{2.24}$$

The graph consists of sets of verticles (nodes) $\mathcal{V}$ related to physical locations, and edges $\mathcal{E} \subset \mathcal{V} \times \mathcal{V}$ which connects pair of verticles and are associated with each edge costs $W_{\mathcal{E}}$. Costs are defined by cost function. They are related to motion distance and, in more advanced cases, with safety, clearance, local vision etc. Thus, path planning is divided into two-steps-solution: graph construction and minimum cost path search. Since motion in $\mathcal{C}$-space assumes workspace with static obstacle, the efficient path planning systems might separate global and local planning. Global planning is associated with general path planning task, while local calculates a short term roadmaps that are based on local sensing and where goals are taken from vecticles in a global graph (Fig. 2.11). Local sensing helps in dealing with unexpected dynamic obstacles which are not included in global maps.

2.4.2 Bug Algorithms

One of the simplest groups of motion planning methods is a well known *Bug algorithms* which is related to reacting behavior [36–40]. Motion here is focused on local sensing, trajectory and map building while the initial localization and the goal are global and already known. An earliest bug algorithm was the one with obstacle sensing only without any pose tracking or map memory, it was called Bug-0. The robot's behavior is focused only on goal reaching, as goal coordinates are known, it's possible to find goal direction and follow it. When an obstacle is discovered the avoidance

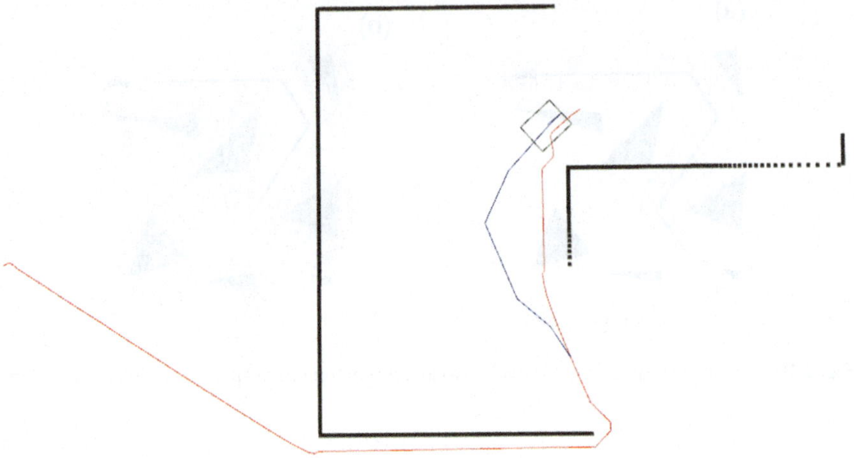

Fig. 2.11 Path planning: global (red) and local (blue) path

direction is drown. The robot follows on obstacle's edge as long as it can return to previous trajectory. The calculated direction trajectory is following until the goal is reached with same handling of any discovered obstacle. This algorithm, enhanced by path memory and simple map building, was introduced as Bug-1. In this version robot handles obstacle by edge following till it backs to the point where the following was started. After that robot chooses the shortest way to point from the completed path, which is the closest to goal and continues moving. The completed path will be suboptimal, however, robot probably, sooner or later, reaches the right place thanks to map information from obstacles it has already met. Improvement of the Bug-1 is called Bug-2, where robot moves toward initial trajectory, called m-line, without any needs for an obstacle edge following. When robot, during boundary-following, encounters m-line, it continues its movement toward the goal. Bug-0/1/2 are mostly based on tactile sensing, and their performance was improved by using range-finder sensor in next Bug algorithms generations, like Dist-Bug and its optimization Intelligent Bug Algorithm (IBA) [37, 40]. The application of range-finders (sonar, IR, laser etc) enables the use of "leaving condition" from the boundary-following step. This condition guarantees that next hit point in observable range will be closer to the goal than in previous case. It means that obstacle avoiding is the most efficient method which cuts resulting completed path in comparison with previous Bugs algorithms. An example trajectories in the same environment are shown on Fig. 2.12.

2.4.3 *Geometry-Based Roadmaps*

Unlike the previous algorithms that do not need map and are based on reacting behavior, the largest group in which the motion planning or roadmaps depend on

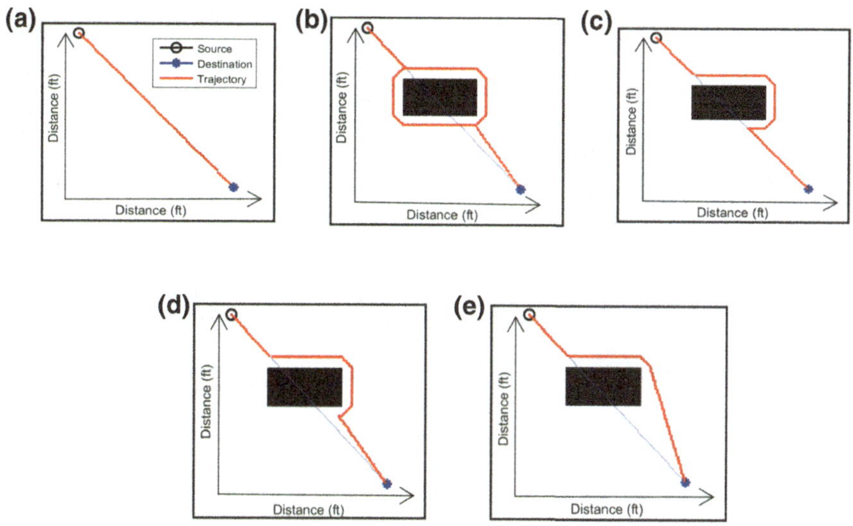

Fig. 2.12 Bug algorithms trajectory comparison: No obstacles (**a**); Bug-1 (**b**); Bug-2 (**c**); Dist-Bug (**d**); IBA (**e**) [40]

word presentation is called *geometry-based roadmaps* and is related to path finding in mentioned graph $\mathcal{G}$ (2.24).

As it was explained in path planning introduction, the two problems of planning that need to be solved can be defined as: graph construction and shortest path search.

The first step in solving the problem of graph construction is to construct a graph which represents the workspace. In Bug algorithms it was allowed to touch the obstacles without penetrating them. The same assumption is used by *visibility graphs* approach for finding a shortest path, but as distinct from bug algorithms this one is based on global sensing and planning [28, 33, 34]. Graph is constructed from visible to each other vertices of polygonal obstacles ($\mathcal{C}_{obs}$) connections or segments. The shortest path will be polygonal which is chosen from these segments (Fig. 2.13a). However, the real shortest path includes obstacle touching, this assumption is not acceptable for many applications as physical contact with obstacles might injure robot's systems and does not have any clearance. More safe path planning approach with maximal clearance assumption is based on *Voronoi diagrams* [28, 41, 42]. Resulting path in this approach is safer but takes much more time than shorter possible path and potentially includes sharper turns (Fig. 2.13b). There are also many hybrid Visibility-Voronoi solution applications that optimize path length and smoothing [43]. There are some weak points of Voronoi diagram approaches such as the hard localization in some cases due to long distance to reference points (obstacles). Another disadvantages of the approach are that dynamic environment causes huge changes in diagrams structure making the system unstable and the generally huge calculation load for non-polygonal and large-dimensioned spaces.

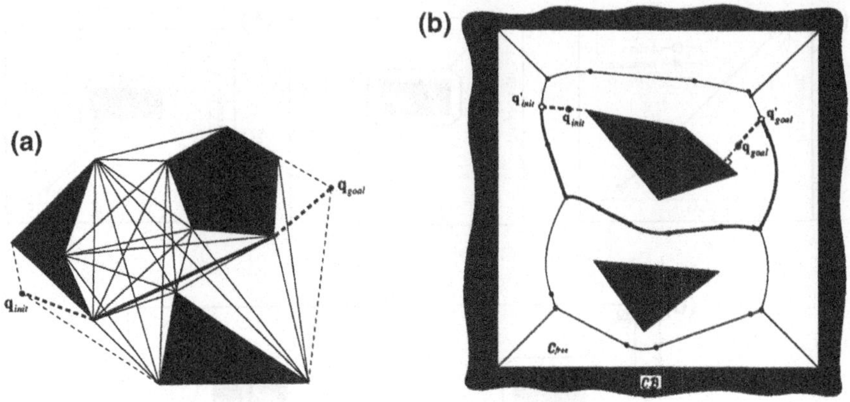

Fig. 2.13 Roadmaps based on visibility graphs (**a**) and Voronoi diagrams (**b**) [28]

Another, geometry-based method that introduces word representation as *exact cells* is *Trapezoidal decomposition* [44]. In this method, space cell decomposition is given by drawing vertical lines from each vertex including outer boundaries to closest edge. It guarantees that any kind of trajectory inside each trapezoidal cell is safe and free from obstacles. Thus, resulting $\mathcal{C}_{free}$ might be interpreted as graph where verticles are trapezoidal cells edge midpoints, connected with themselves if edges belong to the same trapezoid. Then, the shorter way can be calculated and the resulting path can be chosen (Fig. 2.14).

Exact cell decomposition is a significant practical approach that might be also extended to non-polygonal obstacle boundaries [28]. When graph construction is done, the next step is to find the most efficient path through nodes in a graph.

The fundamentals of the shortest path finding was introduced by Dijkstra in 1959 [45]. His algorithm calculates paths from single vertex X_{start} to all vertices $\mathcal{V}$ in a given graph $\mathcal{G}$ by visiting all of them. The cost of each path is computed basing on distances, so Dijkstra's algorithm is usually used in many different path planners where there is a need for choosing the shorter possible path between two vertices, after calculation is finished. This algorithm is also one of those which are used in this book. Dijkstra computes all possible paths which requires a lot of computing time in large graphs while often only path between X_{start} and target is essential. A* algorithm that was developed by Hart et al. in 1968 introduces a heuristic function

$$f(X) = g(X) + h(X) \tag{2.25}$$

to cost estimation in path searching [46]. Where $g(X)$ is related to actual cost and $h(X)$ is the estimated cost, like Euclidean distance from X_{start} target. It allows to search only in nodes (vertices) with minimal costs during expansion without costs calculating for already visited vertices. It means that the path calculation will be as fast as possible until better-informed heuristic is not included. However, since A*

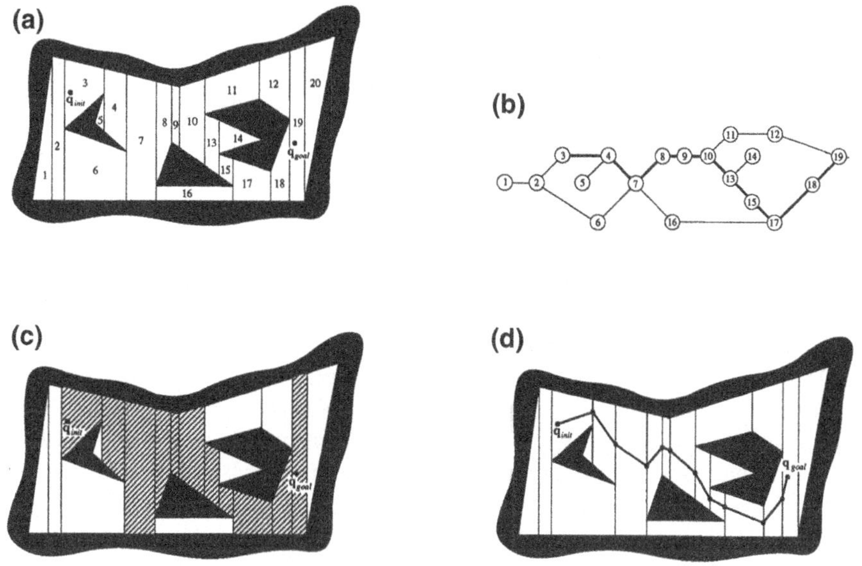

Fig. 2.14 Exact cell (Trapezoidal) decomposition path planning method: $\mathcal{C}_{free}$ cell decomposition (**a**); its graph (**b**); shortest path choice (**c**); and resulting trajectory (**d**) [28]

assumes a static $\mathcal{C}$-space where any changes like new obstacles will be observed it must be re-run from the beginning. The dynamic version of A* named as D* was developed by A. Stentz and first introduced in [47]. Since in dynamic environment A* might be very expensive, D* does not require full re-planning from scratch the costs are changeable locally and incrementally. It makes D* performance in dynamic workspaces comparable to A* in static. There are many different path planning approaches which are based on A* and D* implemented during past decades like Incremental A*, Anytime Dynamic A*, Focussed D*, D* Lite, Delayed D*, Moving Obstacle Avoiding A* etc [48–54]. They might perform more efficient search in applied cases such as dynamic environments or even multirobots planning.

2.4.4 Sampling-Based Roadmaps

Probably the simplest world discretization (or sampling) technique is divided to finite cells called *grids* so it is directly connected to metric maps described in Sect. 2.3.2. Path planning task in this grid map representation is similar to regular one as black cell represents $\mathcal{C}_{obs}$ and trajectory is searching in $\mathcal{C}_{free}$—white cells. As it was already mentioned cells can been seen as graph vertices with finite size. Therefore, after graph construction, the shortest path algorithms can be applied.

A sampling opens door to a *probabilistic roadmap* also known as PRM techniques development. Fundamental work in this field is Kavraki's "Probabilistic roadmaps for path planning in high-dimensional configuration spaces" [55] which had many different extension. The Kavraki assumes that at the initial state there is no information about C_{obs} and C_{free}, the C is discovered by randomly drawn samples (nodes) and tried to be connected nearby by shortest possible path with local planner usage. Obstacle-free connections between nodes build C_{free}. The performance in this solution depends on predefined number of nodes n and their connections k or specified connection radius. Although this approach performs generally well, it seems to be the best for applications with more complex robot configurations. One of most important PRM disadvantages is high susceptibility to missing vertices in narrow passages that could, in some cases, cause an unsolvable matter and results in uncompleted space maps because of disconnected nodes.

The last method presented in this section, is a practical space-filling method—*Rapidly-exploring random trees* or RRT—first introduced by LaValle. It finds wide path planning and exploration usage in high-dimensional C constrained not only from obstacles (as most techniques presented before), but also from nonholonomy and dynamics [56–58]. The basis is an incremental space expansion done by search tree from sampled points, which present a random robot configuration. This method is easy to implement as well as to choose expansion strategy which dependents on the task. This can be a focus on expanding goal direction and finding shortest path instead of random expansion for space exploration. As RRT is not limited to one sampled point at time, in many cases it uses bidirectional search with separate RRTs from initial state and goal that find each other. Similarly to PRM, rapidly-exploring random trees technique finds narrow passages as problematic.

2.4.5 Potential Field Path Planning

Potential fields is an artificial method for a "local" path planning based on incremental graph construction in C_{free} by a step size δ_i, from the local minima of a potential function $\mathbf{U}(X)$ to a goal X_g. This approach was first introduced by Khatib for manipulator trajectory planning and online obstacle avoidance by continuous sensing [59]. The numerical potential field techniques for robot planning are well described by Barraquand et al. in their work [60]. This approach analyze regarding to mobile robot navigation was presented by Koren and Borenstein in [61].

A potential function must be differentiate for every pose $X \in C_{free}$ and could be defined as:

$$\mathbf{U}(X) = \mathbf{U}_{att}(X) + \mathbf{U}_{rep}(X) \tag{2.26}$$

where $\mathbf{U}_{att}(X)$ is named as an *attractive potential* and related to goal pose, and $\mathbf{U}_{rep}(X)$ is a *repulsive potential* which is associated with C_{obs}. The path planning is an interactive process of incremental motion in the direction given by the artificial force $\mathbf{F}(X)$ that is computed for current state X as:

$$\mathbf{F}(X) = -\nabla \mathbf{U}(X) \tag{2.27}$$

The $\Delta\mathbf{U}(X)$ is a gradient of $\mathbf{U}$ at pose X that for 3DOF is described as:

$$\nabla\,\mathbf{U}(X) = \begin{bmatrix} \frac{\partial U}{\partial x} & \frac{\partial U}{\partial y} & \frac{\partial U}{\partial \theta} \end{bmatrix}^T \tag{2.28}$$

Then, after putting (2.27) and (2.28) into an Eq. (2.26), it might have a form:

$$\mathbf{F}(X) = -\,\nabla\,\mathbf{U}_{att}(X) - \nabla\mathbf{U}_{rep}(X) \tag{2.29}$$

The chosen δ_i step must be small enough to find an obstacle avoiding graph segments, but its size decreasing directly the searching time. The step can't also be too big since it may cause goal X_g overshooting and the process could take a large amount of time. One of most widely used approaches is a gradient descent which is also applied for machine learning and neural network methods [62–64]. The potential field path planning approach with its optimization is still of wild research interests among robotics and artificial intelligence communities.

It is easy to find that all already presented navigation and mapping approaches assume that robot localization is precisely known. In the next section the study of several localization techniques is presented.

2.5 Robots Localization

Robot localization, often named as position *estimation* or *tracking* [22], forms a problem of reconstruction robot's pose referencing to given landmarks. Since kinematics in three-dimensional space is described by six coordinates—Cartesian x, y, z and Euler angles ϕ, θ, ψ which might be reduced to less dimensions number, robot's localization or *pose* with respect to used coordinate frame is defined as:

$$\mathbf{X} = \begin{cases} \{x\} & \textbf{for } \textit{one-dimensional space } \mathbf{X}^{(1)} \\ \{x\ y\ \theta\}^T & \textbf{for } \textit{two-dimensional space } \mathbf{X}^{(2)} \\ \{x\ y\ z\ \phi\ \vartheta\ \psi\}^T & \textbf{for } \textit{three-dimensional space } \mathbf{X}^{(3)} \end{cases} \tag{2.30}$$

The localization methods may be classified by two criteria such as indoors or outdoors and local or global. Local methods are mostly related to incremental techniques that are based on local data (encoders, sensors), they give information about object's state and displacements in a local frame. The global methods perform the localization information in global frame. Usually that two techiques are combined and local data is used for global localization calculation. The second classification is based on an target applications that can work inside of isolated man-created environment or natural constructions (indoor) and open terrains (outdoor). In this section we look on different approaches of mobile robots localization and their use cases.

2.5.1 Odometry

Localization based on wheeled *odometry* is probably the oldest, but also the fastest and the easiest short-term method. Despite the fact that there are many various sources of errors that are cumulated in long-term perspective, an odometry still finds wide usage in mobile robotics as a cheep and fast implemented solution. Localization of robot, in this approach is taken from incrementation of passed, in sampling time interval I, distance and rotation that is read from encoders pulses N. For differential drive odometry a traveled distance $\Delta U_{r/l,I}$ is calculated by an equation [65]:

$$\Delta U_{r/l,I} = C_m N_{r/l,I} \tag{2.31}$$

where c_m is a conversion factor that describes linear wheel displacement based on encoder pulses and given from wheels geometry as diameter D_n dependent on encoders resolution C_e and gear ratio factor n:

$$c_m = \pi D_n / n C_e \tag{2.32}$$

Odometry incorrect (unclear) equations for target robotic system could be an additional errors source and must be constructed with required solidity. Due to the wide usage of this technique, Odometry accuracy improvement was a subject of many researches that developed important methods of systematic (for example caused by unequal wheel) and unsystematic (like slippage or other surface-wheel contact unexpected results) errors reducing for single or group of mobile robots [66–70]. Notable systematic errors estimation techniques for differential drive robots is *UMBmark* introduced by Borenstein and Feng [68, 71]. The Kalman filter technique are widely used for odometry calibration, but the efficient usage of odometry data requires vision observation, that causes heavy loading on calculation units [72]. The Kalman Filter basis and localization technique is more detailed described in the next Chap. 3 (Sect. 3.3.3).

2.5.2 Inertial Navigation

Another known incremental localization technique is *Inertial navigation*. The global localization computing for this approach is based on similar to odometry technique—by incrementation of local displacement (passed path) information. Here, sensors are mainly represented by sets of gyroscopes and accelerometers, named as *Inertial measurement units* or IMUs (Fig. 2.15) that are widely used in the inertial navigation systems or simply *INS*. Orientation is given from gyroscopes that provide angular rate information as distinct from an odometry-based approach, where huge orientation errors were observed. It is very precise due to the development of efficient error reducing techniques. Accelerometers presented in IMUs provide displacement

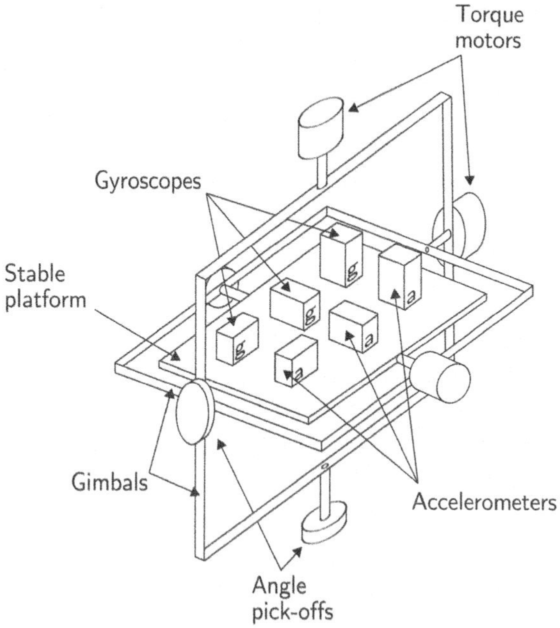

Fig. 2.15 A typical IMU construction [73]

information as a velocity rate, which is given from angular and linear acceleration measurements. These measures are often seen as a weak point of IMU-based applications, since they are a source of many system errors.

Unfortunately, in this approach even small inertial error (like "shifting" in an idle state) can cause a huge global error growth. Therefore, the estimation techniques are already required at early system running states. A simplest error reducing method is a sensor reset after a few rates calculation iterations. This reduces errors cumulation at the beginning and helps to decrease the number of global information faults. However, it is still not enough for the efficient INS usage. An notable accelerometer error estimation technique was introduced by Barshan and Durrant-Whyte in their work [74]. The INS and its errors descriptions in mentioned research might be used as a important background for mobile robots localization systems with a low-cost INS. In this work IMU sensor is used together with GPS and Odometry estimation technique based on an Extended Kalman Filter (EKF). Another interesting work is Woodman's description of Inertial navigation systems that mostly focuses on general, physical and construction concepts [73]. Although, the inertial navigation systems because of their incremental nature are facing with the same localization disaster as an odometry in long-term perspective, the isolation from other systems and environment is their the main advantage. Which together with IMU successive cost decreasing and different calibration techniques development makes this approach is more widely used. And in consequence, the IMU-based localization applications might be found in a large number of mobile robotic researches and systems during last decade [75–78].

2.5.3 Visual and Laser Odometry

Since wheel odometry errors are mostly caused by wheel-surface contact, and robotic systems are usually equipped with vision sensors that could also be used as localization feedback source (see Chap. 3), the idea of *Vision* or *Laser Odometry* (depend of sensor type) was born. Techniques in this group use similar principles as their parent (wheel odometry)—it is incremental and local. The biggest difference and advantage is that movement increments originate from observable landmarks, it means that odometry data is not related to wheels at all. Therefore, it makes visual odometry approach a very strong one for systems where map creating is in a box. Although, the problem of robot's motion estimating from visual sensing dates back to the 80s [79] and is related to a *structure from motion* (SFM) research field [80]. The term Visual Odometry (VO) and one of the first efficient real-time system implementations was introduced in 2004 by Nister et al. in research with same name [81]. Their system was based on a stereo or monocular camera usage and opposed to all previous works, the robot's motion was not calculated from 3D to 3D points but from 3D to 2D. Another visual odometry approach with computing 2D to 2D point with recreating 3D pose was proposed by Comport et al. in [82]. The biggest challenge for all VO techniques is related to the estimated trajectory drift. There are several error reducing approaches being developed that use additional localization estimation techniques with laser, GPS, IMU etc data which allows VO long-term usage in rough terrain [83–85].

The group of approaches based on a very similar to VO concepts but using data from laser scanner or light detection and ranging sensor (LIDAR) is called a Laser Odometry. Distance measurement is based on either 'time of fly' (TOF or 'echo pulse') or phase shifting methods (Fig. 2.16) [86]. TOF uses a very short but intense laser pulse to the object, and after the reflection to the rangefinder. Laser system works on a very precise time measure of interval between the point when laser emitted and received, after the reflection from the measurement object and distance; it is calculated as follows:

$$R = v \cdot t/2 \tag{2.33}$$

where v—speed of electromagnetic radiation and t is measured time interval. In this case, as the speed of light is very accurately known value, the precision of time measurement decides upon laser precision and resolution. The second method uses continuous sinusoid beam which compares transmitted and received wave, and measures phase shifts.

The amplitude of laser radiation is modulated by a sinusoidal signal that has a period T_m and wavelength λ_m. The slant distance R is measured by using measures of phase difference or the phase angle ϕ between emitted beam, and received after reflection from the object. Phase measurement usually uses digital pulse counting techniques that give the fractional part of the total distance ($\Delta\lambda$). The final slant range is provided bythese fractional values, which can be determined from the integer

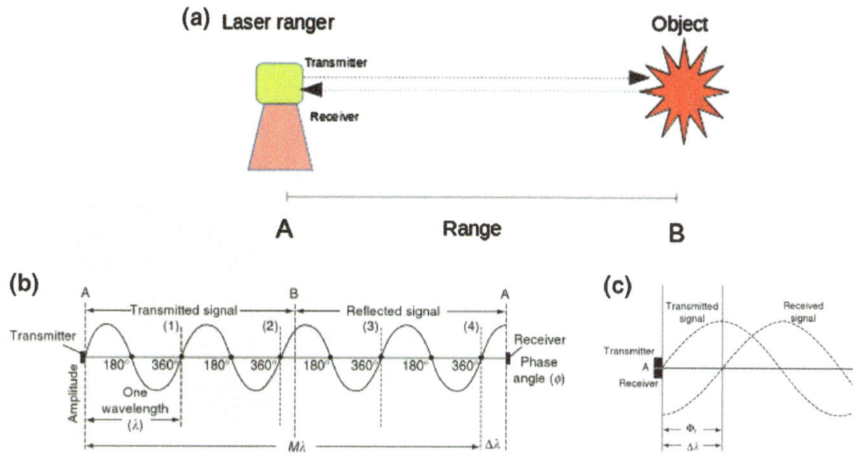

Fig. 2.16 Laser distance measurement methods: 'echo pulse' TOF method (**a**) and phase comparison between transmitted and reflected singals (**b**), and between two signals takes place at the laser rangefinder located at A (**c**)

number of wavelengths (M):

$$R = \frac{M\lambda + \Delta\lambda}{2} \tag{2.34}$$

where M—integer number of wavelengths, λ—known value of wavelengths, $\Delta\lambda = (\frac{\phi}{2\pi})\lambda$—fractional parts of the wavelengths. The velocity of the measured points might be given from Doppler shift effect [87] and for surface matching and pose estimation an Iterative Corresponding Point (ICP) family methods the are often applied [88–91]. One of most robust laser odometry ICP approach is based on point-to-line (PL-ICP) instead of classic in ICP point-to-point technique [92, 93] and uses normals to the surface at the projected points. The exact closest-form solution of lineariazation of non-linear PL-ICP method was introduced as algorithm developed by Censi in [92]. The Efficiency analyzed of this method in comparison to others from ICP family which is based on an experimental data could be found in the same paper as the proposed algorithm. The faster but not so accurate as PL-ICP method is the Polar Scan Matching approach (PSM), which basing on experimental data, is even 6 times faster than vanilla ICP [94]. The combined PL-ICP and PSM algorithm for a walking robot in a structured environment could be found in [95]. This approach uses the PSM algorithm for initial pose estimating and PL-ICP to scan-matching.

The laser approach which is focused on 2D laser odometry estimation in dynamic environment was introduces by Jaimez et al. in paper [96]. This method is named RF2O (Range Flow-based 2D Odometry), is based on measured point motion representation as sensor velocity function with a static word assumption. For applying this technique into dynamic environment, the M-estimator was used. The position of each point related to local laser frame is described by polar coordinates—range r and azimuth θ and measured for each time interval Δt (Fig. 2.17).

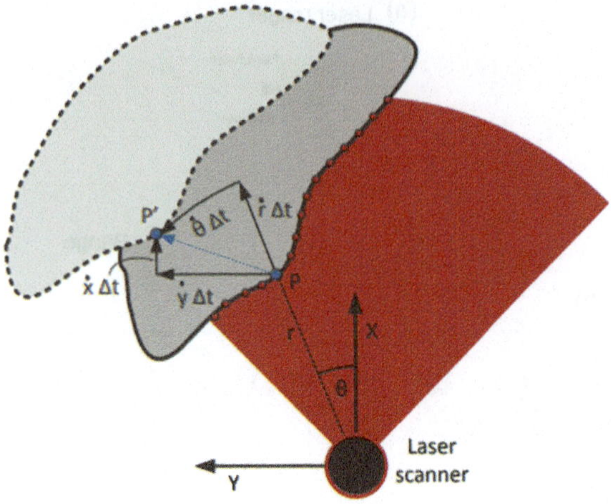

Fig. 2.17 An observable point P motion with respect to lidar frame described by polar coordinates and calculated for each time interval Δt [96]

The transition from polar to Cartesian coordinates by velocities is delivered as:

$$\begin{aligned} \dot{r} &= \dot{x}\cos\theta + \dot{y}\sin\theta \\ r\dot{\theta} &= \dot{y}\cos\theta - \dot{x}\sin\theta \end{aligned} \tag{2.35}$$

And by applying the assumption that the motion of every measured point is related to laser sensor frame, the sensor or robot's velocity can be computed as same but with an opposite sign:

$$\begin{bmatrix} \dot{x} \\ \dot{y} \end{bmatrix} = \begin{bmatrix} -v_{x,s} + y\omega_s \\ -v_{y,s} - x\omega_s \end{bmatrix} \tag{2.36}$$

Then, after connecting (2.35) and (2.36) and applying the estimator, the author finds a robust laser odometry algorithm which according to their paper is more accurate and fast than PL-ICP and PSM. The comparison of these approaches for robot applications in plane and rough terrain cases is also to be found in Chap. 5 of this book.

Another significant pose estimation from laser data approach, named *Laser Odometry and Mapping* (LOAM) was developed by Zhang and Singh [97]. This method uses 3D laser data gathered from 2-axis 2D lidar moving in 6-DOF, and according to the authors doesn't require additional independent localization source for reaching real-time effects comparable to offline systems. However, IMU usage increases accuracy of method in about 2 times (Fig. 2.18).

Loop closure problem that was presented in the early releases of LOAM research has been particularly fixed according to the latest reports, but still has impact on the system for the continuously looping tasks [98]. That report also confirms that laser odometry methods have low-drift and might be used in almost any ground or aerial real-time exploration systems.

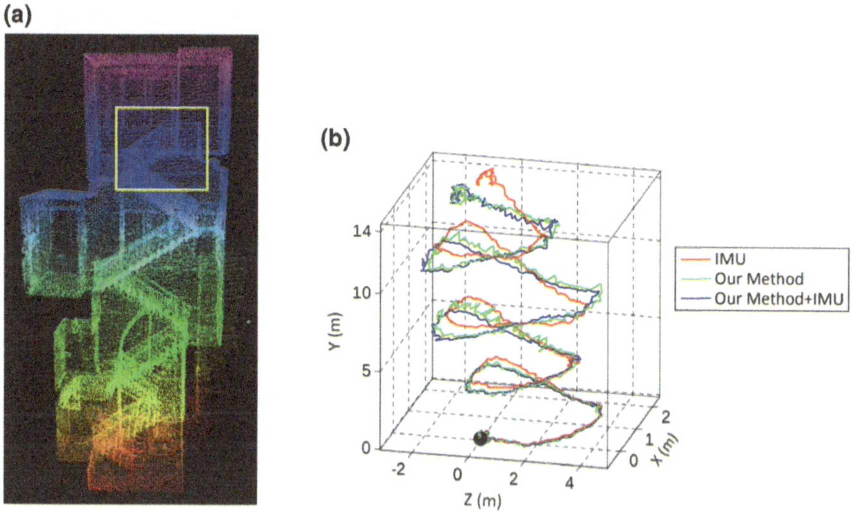

Fig. 2.18 Map created with laser odometry usage (left) and LOAM pose estimation results with/without IMU comparison (right) [97]

2.5.4 *Active Beacon Localization*

Mobile robotics absolute localization methods group, that might be applied for both indoors and outdoors, is called Active Beacons. This approach is based on a set of known beacons with unique ids, that transmit their precise localization. A mobile robot will then use this information to calculate its own pose. There are many different active beacon system solutions proposed [65, 99–101]. The most known ones are *triangulation*—based on three or more active sensors with robot's rotating sensor that measures angles to beacons and calculates x, y coordinates; and a *trilateration*—that uses information from active transmitters placed in a known localization and computes distance to beacon by time of flight data. In this book we will focus more on trilateration methods as some of them can be applied in our researches. Active beacon localization basics might be presented by example of an Active Beacon System (ABS) developed by Yun et al. in [99]. Their robots localization is based on increments and given from mobile robot—beacon RF and ultrasonic receive/transmit communication system (Fig. 2.19).

This approach requires continuous interacting with at least 3–4 active beacon at same time. Localization is calculated as:

$$\begin{bmatrix} d(n+1) \\ \theta(n+1) \end{bmatrix} = \begin{bmatrix} d(n) + \Delta d \\ \theta(n) + \Delta\theta \end{bmatrix} \tag{2.37}$$

where d is a distance to beacon, Δd its change between n and $n+1$ time and θ is an orientation with change $\Delta\theta$.

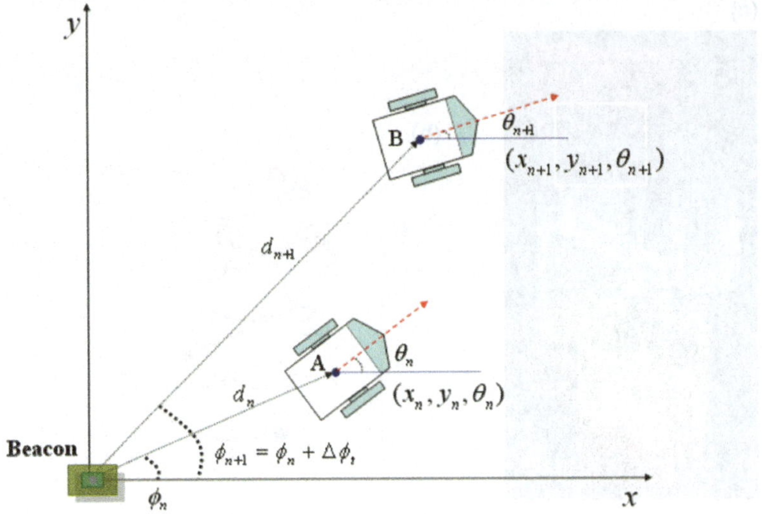

Fig. 2.19 Active beacon linear incremental algorithm [99]

Distance from robot to beacon is given from:

$$d = v \cdot s \tag{2.38}$$

where $s = t_1 - t_0 - t_d$—a time ultrasonic wave traveling, t_0—time when robot sends signal to beacon, t_1—time when robot gives response from a beacon, t_d—delay in time for a beacon; $v = 331.5 + 0.6 \cdot T_a$—wave speed, T_a—room temperature in °C.

Then the robot's coordinates related to world at a time n are being calculated from:

$$\begin{bmatrix} x(n) \\ y(n) \end{bmatrix} = \begin{bmatrix} d \cos\theta(n) \\ d \sin\theta(n) \end{bmatrix} \tag{2.39}$$

Authors also developed Extended Kalman filter pose estimator for their algorithm.

Probably the most known and widely used active beacon trilateration approach is implemented in a *Global positioning system* (GPS). This system has been underdevelopment by the USA Department of Defense for military purpose since 1960s, but found very wide commercial and noncommercial usage including robotics application during the last 30 years. Currently, two separate services for military (a Precise Positioning System or PPS) and civil (Standard Positioning Service or SPS) purposes exist in parallel [102]. Localization on two-dimensional plane xy requires signal from at least three independent sources—satellites with very precise synchronization (atomic clocks). After receiving multiple satellites signal, the equations of position tracking must be solved. Fundamentals and examples might be found in [102–104].

2.6 Summary

The presented chapter gives the background of the mobile robots' fundamentals and a presents a short review of the selected navigation, mapping and localization approaches, that can be taken as an introduction to mobile robotics, as well as give an important background for the integrated approaches development. Robots kinematics and dynamics are an essential parts of the motion controls system implementations. Differential drive mobile robots are non-holonomics systems that need to be considered in path planning, as having on motion control capabilities. Many path-planning approaches assume that low-level controller applied to wheels [...] and motors are constructed well enough for safe motion that excludes major dynamics fails. Therefore, the dynamics models are often not included in planning algorithms. Although, the navigation approaches depend on the type of workspace representation, many of them can be considered as safe path searches from the start to target vertices inside of a graph that represents the operational world. Since the occupancy grids can be seen as graphs of a fixed structure, the shortest path find algorithms like Djikstra's or A* and their successors can be applied in this type of world representation. However, these algorithms generally assume that obstacles are static and any changes in a workspace cause recalculation from scratch. For the workspaces with dynamic obstacles, better solution might be reaction-based path planning approaches. The localization techniques presented in this chapter are not absolute and all of them have both advantages and disadvantages, thus, the efficient robot's localization applications in practice require data fusion or map knowledge. Moreover, the solution of global localization problem is based on map knowledge. Since operational world is dynamic, the map estimation is a separate task that has to be completed. Therefore, it is easy to find that simultaneity of pose and map estimation is an important ability of truly autonomous robots, especially for the inspection applications. This conclusions opens a discussion about estimation techniques which are covered by probabilistic robotics framework and its techniques of simultaneous localization and mapping. The survey of the mentioned approaches will be presented in the next chapter.

References

1. Buratowski, T.: Mobile Robots - Selected Issues. Wydawnictwa AGH, Kraków (2013)
2. Buratowski, T., et al.: A self-stabilising multipurpose single-wheel robot. J. Theor. Appl. Mech. **50**(1), 99–118 (2012)
3. Dudek, G., Jenkin, M.: Computational Principles of Mobile Robotics. Cambridge University Press, Cambridge (2010)
4. Ribeiro, M.I., Lima, P.: Kinematics models of mobile robots (2002). http://users.isr.ist.utl.pt/~mir/cadeiras/robmovel/Kinematics.pdf. Instituto de Sistemas e Robotica. Accessed 12 May 2018
5. Giergiel, M., Hendzel, Z., Ż ylski, W.: Modelowanie i sterowanie mobilnych robotów kołowych. Wydawnictwo Naukowe PWN, Warszawa (2013)
6. Żylski, W.: Kinematyka i dynamika mobilnych robotów kołowych. Rzeszów: OficynaWydawnicza Politechniki Rzeszowskiej (1996)

7. Giergiel, M., Małka, M.: Modelowanie kinematyki i dynamiki mobilnego minirobota. Modelowanie Inżynierskie **1**(32), 157–162 (2006)
8. Burghardt, A.: Modelowanie dynamiki mobilnego robota kołowego równaniami Appella. Acta Mech. Autom. **4**(1), 9–12 (2010)
9. Gutowski, R.: Mechanika analityczna. PaństwoweWydawnictwo Naukowe, Warszawa (1971)
10. Trojnacki, M.: Metody tworzenia modeli dynamiki mobilnych robotów kołowych. Modelowanie Inżynierskie **17**(48), 150–161 (2013)
11. Kane, T.R., Levinson, D.A.: Dynamics, Theory and Applications. McGraw Hill, New York City (1985)
12. Eich-Soellner, E., Führer, F.: Numerical Methods in Multibody Dynamics. Springer, Berlin (1998)
13. Barraquand, J., Latombe, J-C.: Nonholonomic multibody mobile robots: controllability and motion planning in the presence of obstacles. Algorithmica **10**(2–4), 121 (1993)
14. Thrun, S. et al.: Robotic mapping: a survey. In: Exploring artificial intelligence in the new millennium 1, pp. 1–35 (2002)
15. Alberto, P., et al.: Efficient integration of metric and topological maps for directed exploration of unknown environments. Robot. Auton. Syst. **41**(1), 21–39 (2002)
16. Fabrizi, E., Saffiotti, A.: Extracting topology-based maps from gridmaps. In: 2000 Proceedings of the IEEE International Conference on Robotics and Automation (ICRA), vol. 3, pp. 2972–2978. IEEE (2000)
17. Dissanayake, M.G. et al.: A solution to the simultaneous localization and map building (SLAM) problem. IEEE Trans. Robot. Autom. **17**(3), 229–241 (2001)
18. Thrun, S.: Learning metric-topological maps for indoor mobile robot navigation. Artif. Intell. **99**(1), 21–71 (1998)
19. Miejskie Przedsiebiorstwo Komunikacyjne S.A. of Kraków. Bus and tram city routes in Krakow. Available: http://www.mpk.krakow.pl/pl/mapki-komunikacyjne/. 2018 (accessed June 10, 2018)
20. Angeli, A. et al.: Visual topological SLAM and global localization. In: 2009 Proceedings of the IEEE International Conference on Robotics and Automation (ICRA), pp. 4300–4305. IEEE (2009)
21. Aurenhammer, F.: Voronoi diagrams-a survey of a fundamental geometric data structure. ACM Comput. Surv. (CSUR) **23**(3), 345–405 (1991)
22. Thrun, S., Burgard W., Fox, D.: Probabilistic Robotics. The MIT Press, Cambridge (2005)
23. Burgard, W., Hebert, M., Bennewitz, M.: World modeling. In: Siciliano, B., Khatib, O. (eds.) Springer Handbook of Robotics, pp. 1135–1152. Springer, Berlin (2016)
24. Moravec, H.P.: Sensor fusion in certainty grids for mobile robots. AI Mag. **9**(2), 61 (1988)
25. Elfes, A.: Using occupancy grids for mobile robot perception and navigation. Computer **22**(6), 46–57 (1989)
26. Finkel, R.A., Bentley, J.L.: Quad trees a data structure for retrieval on composite keys. Acta Inf. **4**(1), 1–9 (1974)
27. Mehta, D.P., Sahni, S.: Handbook of Data Structures and Applications. Taylor & Francis, London (2018)
28. Latombe, J.C.: Robot Motion Planning. Springer Science & Business Media, Germany (2012)
29. Sünderhauf, N.: Robust optimization for simultaneous localization and mapping. Ph.D. thesis. Technischen Universitat Chemnitz (2012)
30. Guivant, J.E., Masson, F.R., Nebot, E.M.: Simultaneous localization and map building using natural features and absolute information. Robot. Auton. Syst. **40**(2–3), 79–90 (2002)
31. Murphy, R.: Introduction to AI Robotics. MIT press, Cambridge (2000)
32. Siegwart, R., Nourbakhsh, I.R., Scaramuzza, D.: Introduction to Autonomous Mobile Robots. MIT press, Cambridge (2011)
33. LaValle, S.M.: Planning Algorithms. Cambridge University Press, Cambridge (2006)
34. Lozano-Pérez, T., Wesley, M.A.: An algorithm for planning collision-free paths among polyhedral obstacles. Commun. ACM **22**(10), 560–570 (1979)

35. Burgard, W., et al.: Introduction to Mobile Robotics: Robot Motion Planning (2018). http://ais.informatik.uni-freiburg.de/teaching/ss11/robotics/slides/18-robot-motion-planning.pdf. Accessed 18 March 2018
36. Yufka, A., Parlaktuna, O.: Performance comparison of the BUG's algorithms for mobile robots. In: Proceedings of the International Symposium on Innovations in Intelligent Systems and Applications (INISTA'09), pp. 416–421. IEEE (2009)
37. Kamon, I., Rivlin, E.: Sensory-based motion planning with global proofs. IEEE Trans. Robot. Autom. **13**(6), 814–822 (1997)
38. Lumelsky, V.J., Stepanov, A.A.: Path-planning strategies for a point mobile automaton moving amidst unknown obstacles of arbitrary shape. Algorithmica **2**(1-4), 403–430 (1987)
39. Ng, J., Bräunl, T.: Performance comparison of bug navigation algorithms. J. Intell. Rob. Syst. **50**(1), 73–84 (2007)
40. Zohaib, M. et al.: Intelligent bug algorithm (iba): a novel strategy to navigate mobile robots autonomously (2013). arXiv:1312.4552
41. Aurenhammer, F., Edelsbrunner, H.: An optimal algorithm for constructing the weighted Voronoi diagram in the plane. Pattern Recognit. **17**(2), 251–257 (1984)
42. Yap, C.K.: An O(n log n) algorithm for the Voronoi diagram of a set of simple curve segments. Discret. Comput. Geom. **2**(4), 365–393 (1987)
43. Wein, R., Van den Berg, J.P., Halperin, D.: The visibility-Voronoi complex and its applications. Comput. Geom. **36**(1), 66–87 (2007)
44. Chazelle, B.: Approximation and decomposition of shapes. Algorithm. Geom. Aspects Robot. **1**, 145–185 (1985)
45. Dijkstra, E.W.: A note on two problems in connexion with graphs. Numer. Math. **1**(1), 269–271 (1959)
46. Hart, P.E. Nilsson, N.J., Raphael, B.: A formal basis for the heuristic determination of minimum cost paths. IEEE Trans. Syst. Sci. Cybern. **4**(2), 100–107 (1968)
47. Stentz, A.: Optimal and efficient path planning for partially-known environments. In: Proceedings of the IEEE International Conference on Robotics and Automation (ICRA), pp. 3310–3317. IEEE (1994)
48. Stentz, A.: The focussed D* algorithm for real-time replanning. In: Proceedings of the International Joint Conference on Artificial Intelligence (IJCAI), vol. 95, pp. 1652–1659 (1995)
49. Ferguson, D., Stentz, A.: The delayed D* algorithm for efficient path replanning. In: Proceedings of the IEEE International Conference on Robotics and Automation (ICRA), pp. 2045–2050. IEEE (2005)
50. Duchoň, F., et al.: Path planning with modified a star algorithm for a mobile robot. Proc. Eng. **96**, 59–69 (2014)
51. Likhachev, M. et al.: Anytime dynamic a*: An anytime, replanning algorithm. In: Proceedings of the International Conference on Automated Planning and Scheduling (ICAPS), pp. 262–271 (2005)
52. Liu, Q., et al.: Global path planning for autonomous vehicles in off-road environment via an A-star algorithm. Int. J. Veh. Auton. Syst. **13**(4), 330–339 (2017)
53. Koenig, S., Likhachev, M., Incremental a. In: Proceedings of the Advances in Neural Information Processing Systems, pp. 1539–1546 (2002)
54. Wang, Z.: Path planning for first responders in the presence of moving obstacles. Arch. Built Environ. **4**, 80–97 (2015)
55. Kavraki L.E., et al.: Probabilistic roadmaps for path planning in highdimensional configuration spaces. IEEE Trans. Robot. Autom. **12**(4), 566–580 (1996)
56. Kuffner, J.J., LaValle, S.M.: RRT-connect: an efficient approach to single-query path planning. In: Proceedings of the IEEE International Conference on Robotics and Automation (ICRA), vol. 2, pp. 995–1001. IEEE (2000)
57. LaValle, S.M.: Rapidly-exploring random trees: A new tool for path planning. Iowa State University, Technical report, Department of Computer Science (1998)
58. LaValle, S.M., Kuffner, Jr J.J.: Randomized kinodynamic planning. Int. J. Robot. Res. **20**(5), 378–400 (2001)

59. Khatib, O.: Real-time obstacle avoidance for manipulators and mobile robots. In: Cox J.C., Wilfond, G.T. (eds.) Autonomous Robot Vehicles, pp. 396–404. Springer (1986)
60. Barraquand, J., Langlois, B., Latombe, J-C.: Numerical potential field techniques for robot path planning. IEEE Trans. Syst. Man Cybern. **22**(2), 224–241 (1992)
61. Koren, Y., Borenstein, J.: Potential field methods and their inherent limitations for mobile robot navigation. In: Proceedings of the IEEE International Conference on Robotics and Automation, pp. 1398-1404. IEEE (1991)
62. Ratliff, N. et al.: CHOMP: gradient optimization techniques for efficient motion planning. In: Proceedings of the IEEE International Conference on Robotics and Automation (ICRA), pp. 489–494. IEEE (2009)
63. Vadakkepat, P., Tan, K.C., Ming-Liang, W.: Evolutionary artificial potential fields and their application in real time robot path planning. In: Proceedings of the Evolutionary Computation Congress, vol. 1, pp. 256–263. IEEE (2000)
64. Yang, S.X., Luo, C.: A neural network approach to complete coverage path planning. IEEE Trans. Syst. Man Cybern. Part B (Cybern.) **34**(1), 718–724 (2004)
65. Borenstein, J. et al.: Mobile robot positioning-sensors and techniques. Technical report, San Diego CA, US: Naval command control, ocean surveillance center RDT, and E DIV (1997)
66. Martinelli, A.: Evaluating the odometry error of a mobile robot. In: Proceedings of the IEEE/RSJ International Conference on Intelligent Robots and Systems (2002)
67. Martinelli, A. et al.: Simultaneous localization and odometry calibration for mobile robot. In: Proceedings of the IEEE/RSJ International Conference on Intelligent Robots and Systems (IROS), vol. 2, pp. 1499–1504. IEEE (2003)
68. Borenstein, J., Feng, L.: Measurement and correction of systematic odometry errors in mobile robots. IEEE Trans. Robot. Autom. **12**(6), 869–880 (1996)
69. Rekleitis, I.M., Dudek, G., Milios, E.E.: Multi-robot exploration of an unknown environment, efficiently reducing the odometry error. In: Proceedings of the International Joint Conference on Artificial Intelligence (IJCAI), vol. 2, pp. 1340–1345 (1997)
70. Buratowski, T., et al.: The precise odometry navigation for the group of robots. Schedae Inf. **19**, 99–111 (2015)
71. Borenstein, J., Feng, L.: UMBmark: a benchmark test for measuring odometry errors in mobile robots. In: International Society for Optics and Photonics Proceedings of Mobile Robots X, vol. 2591, pp. 113–125 (1995)
72. Larsen, T.D. et al.: Location estimation for an autonomously guided vehicle using an augmented kalman filter to autocalibrate the odometry. In: Proceedings of the 1st International Conference on Multisource-Multisensor Information Fusion, FUSION98 (1998)
73. Woodman, O.J.: An introduction to inertial navigation. Technical report, University of Cambridge, Computer Laboratory, Cambridge, UK (2007)
74. Barshan, B., Durrant-Whyte, H.F. Inertial navigation systems for mobile robots. In: IEEE Trans. Robot. Autom. **11**(3), 328–342 (1995)
75. Simanek, J., Reinstein, M., Kubelka, V.: Evaluation of the EKF-based estimation architectures for data fusion in mobile robots. IEEE/ASME Trans. Mechatron. **20**(2), 985–990 (2015)
76. Yi, J. et al.: IMU-based localization and slip estimation for skid-steered mobile robots. In: Proceedings of IEEE/RSJ International Conference on Intelligent Robots and Systems (IROS), pp. 2845–2850. IEEE (2007)
77. Ning, M., Zhang, S., Wang, S.: A Non-cooperative satellite feature point selection method for vision-based navigation system. Sensors **18**(3), 854 (2018)
78. Malyavej, V., Kumkeaw, W., Aorpimai, M.: Indoor robot localization by RSSI/IMU sensor fusion. In: 10th International Conference on Proceedings of the Electrical Engineering/Electronics, Computer, Telecommunications and Information Technology (ECTI-CON), pp. 1–6. IEEE (2013)
79. Moravec, H.P.: Obstacle avoidance and navigation in the real world by a seeing robot rover. Ph.D. thesis. Stanford University (1980)
80. Fraundorfer, F., Scaramuzza, D.: Visual odometry: Part i: the first 30 years and fundamentals. IEEE Robot. Autom. Mag. **18**(4), 80–92 (2011)

81. Nistér, D., Naroditsky, O., Bergen, J.: Visual odometry. In: Proceedings of the IEEE Computer Society Conference on Computer Vision and Pattern Recognition (CVPR), vol. 1, pp. 964–971. IEEE (2004)
82. Comport, A.I., Malis, E., Rives, P.: Accurate quadrifocal tracking for robust 3d visual odometry. In: Proceedings of the IEEE International Conference on Robotics and Automation, pp. 40–45. IEEE (2007)
83. Mourikis, A.I., Roumeliotis, S.I.: A multi-state constraint Kalman filter for vision-aided inertial navigation. In: Proceedings of the IEEE International Conference on Robotics and automation, pp. 3565–3572. IEEE (2007)
84. Konolige, K., Agrawal, M., Sola, J.: Large-scale visual odometry for rough terrain. In: Robotics Research, pp. 201–212 (2010). Springer
85. Jones, E.S., Soatto, S.: Visual-inertial navigation, mapping and localization: a scalable real-time causal approach. Int. J. Robot. Res. **30**(4), 407–430 (2011)
86. Heritage, G., Large, A.: Laser Scanning for the Environmental Sciences, pp. 3–6. Wiley, New York City (2009)
87. Horn, J., Schmidt, G.: Continuous localization of a mobile robot based on 3D-laser-range-data, predicted sensor images, and dead-reckoning. Robot. Auton. Syst. **14**(2-3), 99–118 (1995)
88. Lu, F., Milios, E.: Robot pose estimation in unknown environments by matching 2d range scans. J. Intell. Robot. Syst. **18**(3), 249–275 (1997)
89. Minguez, J., Montesano, L., Lamiraux, F.: Metric-based iterative closest point scan matching for sensor displacement estimation. IEEE Trans. Robot. **22**(5), 1047–1054 (2006)
90. Haralick, R.M. et al.: Pose estimation from corresponding point data. In: Freeman, H. (ed.) Machine Vision for Inspection and Measurement, pp. 1–84. Elsevier (1989)
91. Rusinkiewicz, S., Levoy, M.: Efficient variants of the ICP algorithm. In: Proceedings of the Third International Conference on 3-D Digital Imaging and Modeling, pp. 145–152. IEEE (2001)
92. Censi, A.: An ICP variant using a point-to-line metric. In: Proceedings of the IEEE International Conference on Robotics and Automation (ICRA), pp. 19-25. IEEE (2008)
93. Besl, P.J., McKay, N.D.: Method for registration of 3-D shapes. In: Proceedings of the Sensor Fusion IV: Control Paradigms and Data Structures, vol. 1611. International Society for Optics and Photonics, pp. 586–607 (1992)
94. Diosi, A., Kleeman, L.: Fast laser scan matching using polar coordinates. Int. J. Robot. Res. **26**(10), 1125–1153 (2007)
95. Skrzypczynski P.: Laser scan matching for self-localization of a walking robot in man-made environments. Ind. Robot: Int. J. **39**(3), 242–250 (2012)
96. Jaimez, M., Monroy, J., Gonzalez-Jimenez, J.: Planar odometry from a radial laser scanner. A range flow-based approach. In: 2016 IEEE International Conference on Proceedings of the Robotics and Automation (ICRA), pp. 4479–4485. IEEE, Stockholm, Sweden (2016). https://doi.org/10.1109/ICRA.2016.7487647, http://mapir.isa.uma.es/mapirwebsite/index.php/mapir-downloads/papers/217
97. Zhang, J., Singh, S.: LOAM: lidar odometry and mapping in realtime. In: Proceedings of the Robotics: Science and Systems 2014, vol. 7. University of California, Berkeley (2014)
98. Zhang, J., Singh, S.: Low-drift and real-time lidar odometry and mapping. Auton. Robots **41**(2), 401–416 (2017)
99. Yun, J., Kim, S., Lee, J.: Robust positioning a mobile robot with active beacon sensors. In: Proceedings of the International Conference on Knowledge-Based and Intelligent Information and Engineering Systems, pp. 890-897. Springer (2006)
100. Peca, M.: Ultrasonic localization of mobile robot using active beacons and code correlation. In: Proceedings of the International Conference on Research and Education in Robotics, pp. 116–130. Springer (2009)
101. Kim, S., et al.: Dynamic localization of a mobile robot with active beacon sensors. IFAC Proc. Vol. **39**(16), 921–925 (2006)
102. Kaplan, E., Hegarty, C.: Understanding GPS: Principles and Applications. Artech house, Norwood (2005)

103. Blewitt, G.: Basics of the GPS technique: observation equations. In: Geodetic Applications of GPS, pp. 10–54 (1997)
104. Bancroft, S.: An algebraic solution of the GPS equations. IEEE Trans. Aerosp. Electron. Syst. **1**, 56–59 (1985)

Chapter 3
SLAM as Probabilistic Robotics Framework Approach

Abstract In this chapter, SLAM is considered as a probabilistic approach that originates from Bayes rule and Markov assumption. The estimation techniques for the robot's pose and map are presented as parts of a probabilistic framework. The Bayesian recursive estimation definition and the Bayes filter implementations, like Kalman and discrete particle filters, were studied here. The map estimation is based on three- and two-dimensional occupancy grids. Last part of this chapter connects these two estimation values into a common one technique that performs they simultaneously. Three different SLAM paradigms—EKF-based, particle-based and Graph-based or Topographic SLAM—have been discussed.

Keywords SLAM · Particle filter · EKF · Probabilistic robotics · Localization · Occupancy grids

3.1 Introduction

This chapter is dedicated to probabilistic framework—a set of approaches in mobile robotics that is based on the probabilistic theory. Chapter includes the introduction of Bayes and Markov rules that are widely used in the framework; overview and discussion of different approaches to probabilistic mobile robots control developed by the robotic and artificial intelligence communities and analysis of works relevant to the book.

It starts from the basic definition of the motion and the observation used in these approaches, and a brief overview of the mathematical tools for robot's pose estimation with a more detailed discussion of Markov localization family including Particle filtering like Adaptive Monte-Carlo Localization and Gaussian like Kalman Filter.

A separate part of this book is dedicated to terrain mapping with landmarks. The knowledge, acquired in the previous chapter about mapping is completed with a more

A. Kudriashov et al., *SLAM Techniques Application for Mobile Robot in Rough Terrain*, Mechanisms and Machine Science 87, https://doi.org/10.1007/978-3-030-48981-6_3

specific focus on the occupancy grid maps and their three-dimensional representation by octrees.

Finally, in last part of the book the learned information are put together to introduce and discuss simultaneous localization and mapping process (SLAM). After defining general SLAM problem and their two typical solutions—smoothing and filtering, three different paradigms—Gaussian (Kalman), Particle filter sampling, with the focus on Rao–Blackwellized approach, and Graph-Based SLAM are discussed.

Analyzing the presented in this chapter are necessary in finding the solution to the book problem in the next parts of the research.

3.2 Motion and Observation Models

One of most important aspects for precise localization and mapping within probabilistic framework, is the description of robot's motion and observation by system specific models called a *motion model* and a *observation model*. Before we describe them, the basic probabilistic definitions and notation need to be defined.

The probabilistic approach uses several different pose X definitions that describe the probability of robot's localization from given values depending on such parameters like map, controls, observations etc:

$$p(X|m) \tag{3.1}$$

$$p(X_t|U_t, X_{t-1}) \tag{3.2}$$

$$p(X_t|Z_t) \tag{3.3}$$

Observations Z for each time from 0 to t are defined by the vector:

$$Z_{0:t} = \left\{ Z_0 \ Z_1 \ \cdots \ Z_t \right\} \tag{3.4}$$

Controls U in the same way is a vector:

$$U_{0:t} = \left\{ U_0 \ U_1 \ \cdots \ U_t \right\} \tag{3.5}$$

3.2.1 Motion Model

Equation (3.2) is defining probabilistic kinematic or *motion model* which will be described in a detailed way in this part. According to many authors motion models

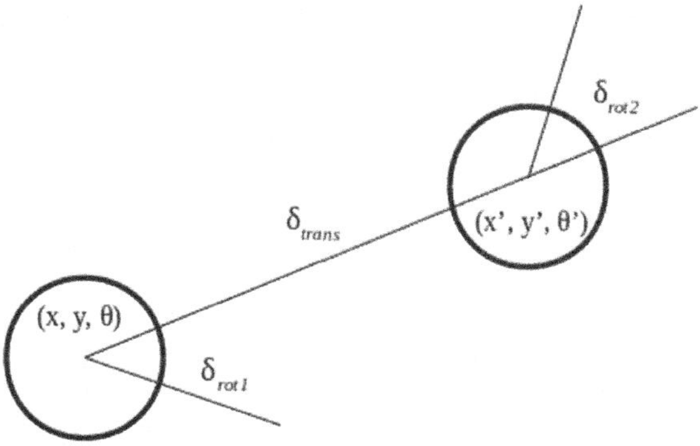

Fig. 3.1 Motion odometry-based model state transition

could be divided into two main groups: *velocity-based* (or dead reckoning) and *odometry-based* [1].

An odometry-based motion model is based, as its name suggests, on data from wheel encoders with possible disturbances and errors specific to wheels odometry. In two-dimensional space transition from $X_{t-1} = \{x, y, \theta\}$ to $X_t = \{x', y', \theta'\}$ (see Fig. 3.1) is given as translation and rotations that describes controls U:

$$U = \left[\delta_{rot_1}, \delta_{trans}, \delta_{rot_2}\right] \tag{3.6}$$

Their values are calculated from the following geometric relations:

$$\begin{aligned} \delta_{trans} &= \sqrt{(x' - x)^2 + (y' - y)^2} \\ \delta_{rot_1} &= \arctan 2(y - y, x - x) - \theta \\ \delta_{rot_2} &= \theta' - \theta - \delta_{rot_1} \end{aligned} \tag{3.7}$$

As system is corrupted by motion errors, usually assuming that it is an independent noise which could be described with a normal distribution by zero-mean values ε_b with variance b. Then, 'real' transition values in a noise model are defined with robot specific error parameters β as:

$$\begin{aligned} \hat{\delta}_{trans} &= \delta_{trans} + \varepsilon_{\beta_3|\delta_{trans}|+\beta_3|\delta_{rot_1}+\delta_{rot_2}|} \\ \hat{\delta}_{rot_1} &= \delta_{rot_1} + \varepsilon_{\beta_1|\delta_{rot_1}|+\beta_2|\delta_{trans}|} \\ \hat{\delta}_{rot_2} &= \delta_{rot_2} + \varepsilon_{\beta_1|\delta_{rot_2}|+\beta_2|\delta_{trans}|} \end{aligned} \tag{3.8}$$

So, the pose X_t from X_{t-1} in odometry-base motion model might be described as following:

$$\begin{bmatrix} x' \\ y' \\ \theta' \end{bmatrix} = \begin{bmatrix} x \\ y \\ \theta \end{bmatrix} + \begin{bmatrix} \hat{\delta}_{trans}\cos(\theta + \hat{\delta}_{rot_1}) \\ \hat{\delta}_{trans}\sin(\theta + \hat{\delta}_{rot_1}) \\ \theta + \hat{\delta}_{rot_1} + \hat{\delta}_{rot_2} \end{bmatrix} \tag{3.9}$$

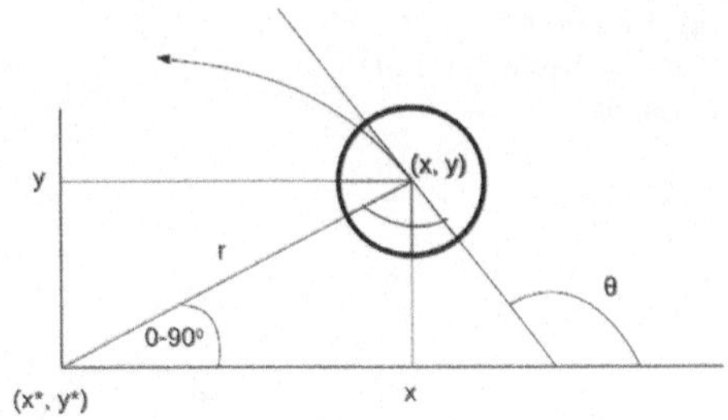

Fig. 3.2 Motion velocity-based model

As opposed to the odometry-based, the *velocity-based motion* model is completely independent from the wheels with their errors, and it assumes that controls are provided only by linear and angular robot's velocities and sample time t:

$$U_t = \begin{bmatrix} V_t \\ \omega_t \end{bmatrix} \tag{3.10}$$

If V and ω are constant during $t-1$ and t robot moves around center point x^*, y^* with $r = |V/\omega|$ (Fig. 3.2).

Relations between central point and coordinates at X_{t_1} for determined time could be described by:

$$\begin{aligned} x^* &= x - \tfrac{V}{\omega} \sin\theta \\ y^* &= y + \tfrac{V}{\omega} \cos\theta \end{aligned} \tag{3.11}$$

Thus, after this determined time denoted as Δt, the X_t coordinates are being given from:

$$\begin{bmatrix} x' \\ y' \\ \theta' \end{bmatrix} = \begin{bmatrix} x^* + \frac{V}{\omega}\sin(\theta + \omega\Delta t) \\ y^* - \frac{V}{\omega}\cos(\theta + \omega\Delta t) \\ \theta + \omega\Delta t \end{bmatrix} = \begin{bmatrix} x \\ y \\ \theta \end{bmatrix} + \begin{bmatrix} -\frac{V}{\omega}\sin(\theta) + \frac{V}{\omega}\sin(\theta + \omega\Delta t) \\ \frac{V}{\omega}\cos(\theta) - \frac{V}{\omega}\cos(\theta + \omega\Delta t) \\ \omega\Delta t \end{bmatrix} \tag{3.12}$$

As the previous approach (3.12) is related to noiseless system, the real probabilistic estimate of velocities is defined with zero-mean error variable ε_b with variance b:

$$\begin{bmatrix} \hat{V} \\ \hat{\omega} \end{bmatrix} = \begin{bmatrix} V \\ \omega \end{bmatrix} + \begin{bmatrix} \varepsilon_{\beta_1|V|+\beta_2|\omega|} \\ \varepsilon_{\beta_3|V|+\beta_4|\omega|} \end{bmatrix} \tag{3.13}$$

Then putting (3.13) into velocity-based motion model (3.12) gives us a real motion model described by velocities in the probabilistic approach:

$$X_t = \begin{bmatrix} x' \\ y' \\ \theta' \end{bmatrix} = \begin{bmatrix} x \\ y \\ \theta \end{bmatrix} + \begin{bmatrix} -\frac{\hat{V}}{\hat{\omega}}\sin(\theta) + \frac{\hat{V}}{\hat{\omega}}\sin(\theta + \hat{\omega}\Delta t) \\ \frac{\hat{V}}{\hat{\omega}}\cos(\theta) - \frac{\hat{V}}{\hat{\omega}}\cos(\theta + \hat{\omega}\Delta t) \\ \hat{\omega}\Delta t \end{bmatrix} \tag{3.14}$$

3.2.2 Observation Model

Another important model is the Observation model which depends on the sensor system and is related to probability of observations from some pose at time t and given map m (3.17).

In general, the observation model might be described by distance (range) and bearing, or just bearing, or distance between the observer and the target. These parameters are to be found in the following equations:

$$d_{ot} = \sqrt{(x_t - x_o)^2 + (y_t - y_o)^2} \tag{3.15}$$

$$\varphi_{ot} = \arctan 2(y_t - y_o, x_t - x_o) - \theta_o \tag{3.16}$$

where $d_o t$ is the distance between observer and target, φ_{ot} is an angle between observer, direction and line that connects the observer with the target, θ_o—the observer's orientation, x_o, y_o, x_t, y_t are the coordinates of the observer and the target. It is easy to explore that more accurate but also more expansive models are based on both bearing and range (Fig. 3.3).

In probabilistic approach, if observations in time from 0 to t are defined as vector (3.4), their probability from localization X inside a known map m at time t might be described as:

$$p(Z_t|X_t, m) \tag{3.17}$$

Each scan Z_t consists of k measurements:

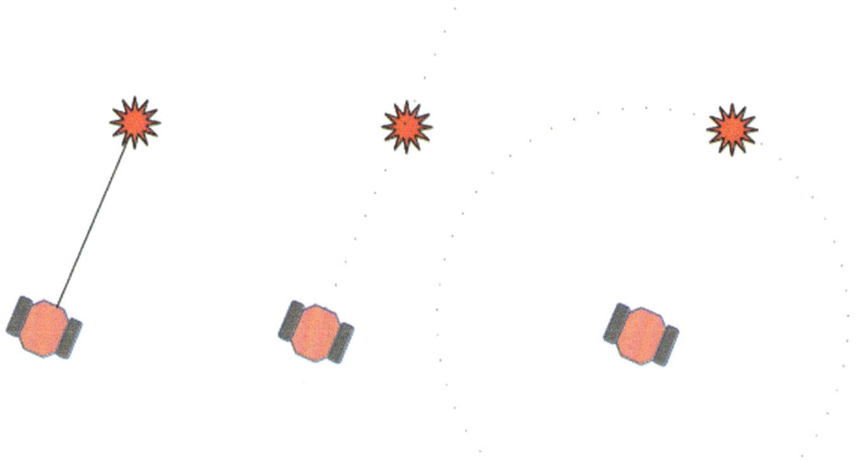

Fig. 3.3 Models of observation due to sensor characteristics—range and bearing (left), bearing only (middle), range only (right)

$$Z_t = \{Z_t^1, Z_t^2, \ldots, Z_t^k\} \tag{3.18}$$

When these measurements could be placed into (3.17), the would be representing the following:

$$p(Z_t|X_t, m) = \prod_{i=1}^{k} p(Z_t^i|X_t, m) \tag{3.19}$$

The Eq. (3.19) is used as the basis for many measurement models, like *laser finder beam* [2, 3] or *likelihood field* [4–6] that is efficiently used in probabilistic maps' applications. The idea of beam model is defined (3.19) as a sum of four distributions—P_{hit}, P_{unexp}, P_{rand} and P_{max} that relate to beams reflected from obstacles, dynamic environment (like moving people), random measurements and measurements related to sensor maximum range:

$$\begin{aligned}
P_{hit}(Z_t|X_t, m) &= \alpha \frac{1}{\sqrt{2\pi b}} \exp^{-\frac{1}{2}\frac{(Z-Z_{exp})^2}{b}} \\
P_{unexp}(Z_t|X_t, m) &= \begin{cases} \alpha\lambda \exp^{-\lambda Z} & Z < Z_{exp} \\ 0 & \text{otherwise} \end{cases} \\
P_{rand}(Z_t|X_t, m) &= \alpha \frac{1}{Z_{max}} \\
P_{max}(Z_t|X_t, m) &= \alpha \frac{1}{Z_{small}}
\end{aligned} \tag{3.20}$$

Therefore, rangefinder beam model could be described as:

$$p(Z_t|X_t, m) = \begin{bmatrix} \beta_{hit} \\ \beta_{unexp} \\ \beta_{rand} \\ \beta_{max} \end{bmatrix}^T \begin{bmatrix} P_{hit}(Z_t|X_t, m) \\ P_{unexp}(Z_t|X_t, m) \\ P_{rand}(Z_t|X_t, m) \\ P_{max}(Z_t|X_t, m) \end{bmatrix} \tag{3.21}$$

And for this approach the equation $P_{hit} + P_{unexp} + P_{rand} + P_{max} = 1$ must be accurate.

Unlike following along the beam, *likelihood model* also called as *scan-based model*, the end point needs to be checked. Thus, observation at time t is represented by:

$$p(Z_t|X_t, m) = \frac{1}{\sqrt{2\pi\sigma}} \exp\left(-\frac{d^2}{2\sigma^2}\right) \tag{3.22}$$

Where d—distance from end-point to the nearest obstacle, that might be calculated in the same way as in (3.15).

3.3 Localization Estimation

Localization estimation is a group of techniques that use probabilistic approach, which is mostly based on Bayes rule, for finding precise localization from given observations and applied controls [1]. In the current section, the fundamentals of Bayesian filtering and selected estimation techniques will be discussed.

3.3.1 *Markov Localization*

According to Bayes Rule, the problem of robot's pose X at each moment of time from 0 to t, basing on observations Z can be described as recursive estimation

$$p(X_{0:t}|Z_{0:t}) = \frac{p(Z_{1:t}|X_{0:t})p(X_{0:t}|Z_{1:t-1})}{p(Z_{1:t}|Z_{1:t-1})} \tag{3.23}$$

Markov localization is a probabilistic method based on Bayes filter that uses observation together with controls for robot's global position estimation [1, 7, 8]. For this model, assuming that the pose of the robot is not exactly known and belief of the robot is under localization x at time t for given observation Z_t is used:

$$Bel(X_t) = p(X_t|Z_t) \tag{3.24}$$

Before the presenting of mathematical derivation will be started, let us look at Markov localization algorithm following the example originally presented by Thrun and Fox.

Robot moves in one-dimensional space on x-axle, that means robot's pose is defined only by x position: $X = [x]$. For the initial state, before any observations are given, $Bel(X_0)$ is distributed with same probability for all locations, meaning the robot is under uncertain state and believes to be everywhere (Fig. 3.4). Based on the observations, the density of probability is changed to a new distribution, which is based on it. As the robot knows that he is close to the door and the sensor systems informs that there are three doors being observed, we can see three peaks of $Bel(X)$ with the same density (Fig. 3.5).

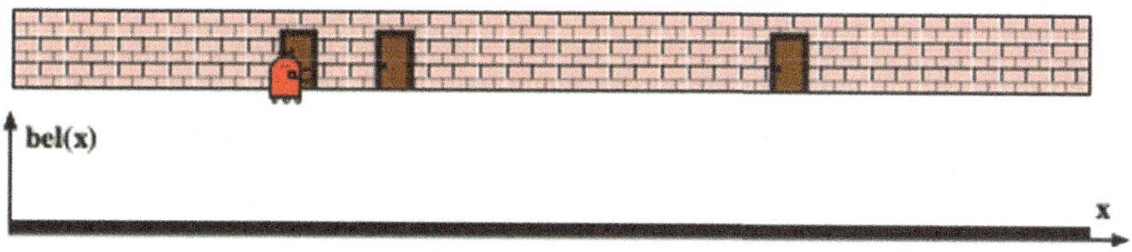

Fig. 3.4 Robot's localization belief at initial state [1]

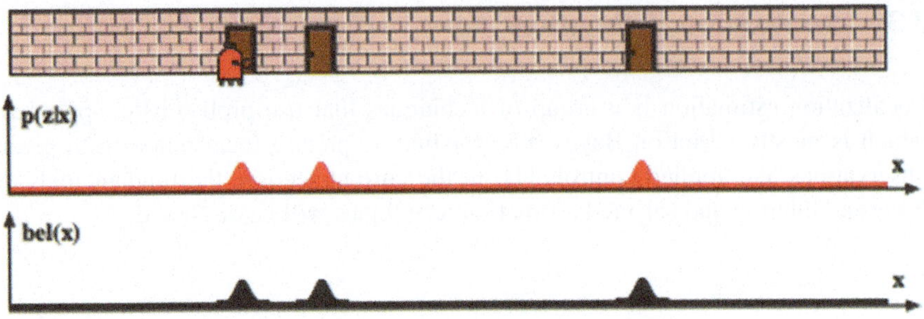

Fig. 3.5 Robot's localization belief after first observation is given [1]

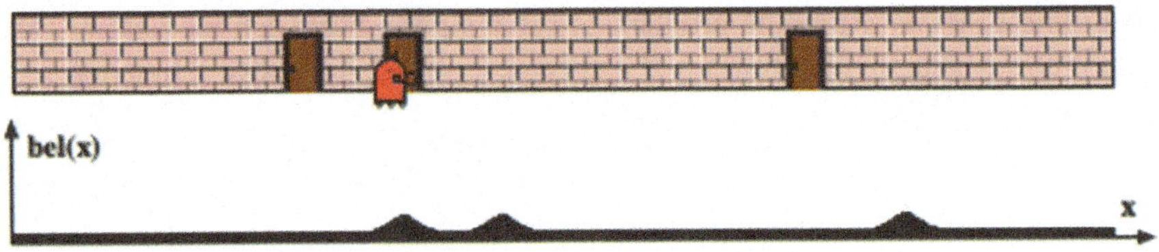

Fig. 3.6 Robot's localization belief after first observation and control are given [1]

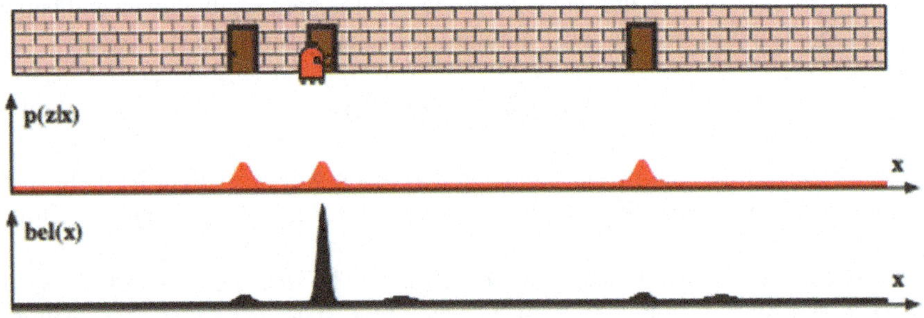

Fig. 3.7 Robot's localization belief after next observation [1]

Now, let us set controls U_t to move the robot forward. This shifts belief distribution since Markov localization assumes that the robot moves according to controls, which impact the whole belief (Fig. 3.6).

Now we need to check if the passed path is in accordance to control, and how the observable environment is changing; this can be checked by the new observations information the was observed. Let us assume that the robot is next to door again, but there are only two doors in front of it that could be observed. The distribution is mostly concentrated at a single location now, the robot assumes with a high rate that it is next to the second door (Fig. 3.7).

Then we add one more control—moving forward to have the second door behind, but enough to reach the third door. Belief in Markov localization calculation will now look like the one in this Fig. 3.8.

Fig. 3.8 Robot's localization belief after final control [1]

So, after few steps with data actualization, it is possible to find the robot's localization by most dens probability field.

Computing of Markov localization might be done by *recursive state estimation* with Bayes filter algorithm as:

$$\underbrace{Bel(X_t)}_{\text{posterior}} = \alpha \cdot \underbrace{p(Z_t \mid X_t)}_{\text{measurement probability}} \cdot \overbrace{\int \underbrace{p(X_t \mid X_{t-1}, U_{t-1})}_{\text{state transition}} \underbrace{Bel(X_{t-1})}_{\text{prior}} \, dX_{t-1}}^{\text{prediction}} \quad (3.25)$$

This algorithm might be divided into two parts: *belief prediction* that use controls (3.26) and *belief distribution* which use the observation (3.27). These steps are also called *prediction* and *correction*:

$$\overline{Bel}(X_t) = \int p(X_t \mid U_t, X_{t-1}, m) \, Bel(X_{t-1}) \, dX_{t-1} \quad (3.26)$$

$$Bel(X_t) = \alpha \, p(Z_t \mid X_t, m) \, \overline{Bel}(X_t) \quad (3.27)$$

This is done recursively for each X_t with computing each $Bel(X_t)$, α as a normalizing constant.

It is also easy to find that the Eqs. (3.26) and (3.27) directly depends on the motion and observation models discussed in Sect. 3.2.

3.3.2 Particle Filters

Particle filters are part of Markov localization family based on *discrete* belief, and together with Gaussian filters are most frequently used Bayesian Filters algorithms. The posterior belief $Bel(X_t)$ is approximated here by a finite number of parameters and particle filter constructs $Bel(X_t)$ recursively from its previous state $Bel(X_{t-1})$. The name *particle* suggests that $Bel(X_t)$ is represented by the "particles"—a set of M samples for each pose state $X^{(i)}$ with the importance factor $w^{(i)}$:

$$Bel(X_t) = \{X_t^{(i)}, w_t^{(i)}\}_{i=1,\dots,M} \tag{3.28}$$

Importance or *weight* factor is weight value of each sample that uses to choose most valued particles in the process. At the *initial* state when the robot's pose is uncertain and represented as uniform distribution, weight value of all samples described as the uniform importance factor $\frac{1}{M}$.

There are different techniques of particle filter implementation, however we will focus on the one called the *adaptive*. This means that the filter can adapt the number of parameters to represent the posterior online. Probably, the most important for this book, is the particle filter algorithm implementation *Adaptive Monte Carlo localization* (AMCL).

The classical example of particle filter implementation is the solution of (3.25), that was introduced by Thrun et al. [9] and Fox et al. [10]. Authors were inspired by Kitagawa implementation of particle filter which is based on Monte Carlo algorithm proposed in [11].

Calculation is divided into *three* phases: *prediction*, *propose* and *correction* from the right to the left side of the Eq. (3.25).

In prediction step a state X_{t-1} is sampled from $Bel(X_{t-1})$ by $X_{t-1}^{(i)}$ from the sample set representing $Bel(X_{t-1})$ with the importance factor $w_{t-1}^{(i)}$.

In propose step a state $X_t^{(i)}$ is sampled by using $X_{t-1}^{(i)}$ sample and controls U_{t-1} from $p(X_t \mid X_{t-1}, U_{t-1})$. The distribution of $X_t^{(i)}$ and $X_{t-1}^{(i)}$ is given by the product $p(X_t \mid X_{t-1}, U_{t-1})\, Bel(X_{t-1})$ and named *proposal distribution.*

In correction step a sample $X_t^{(i)}$ is weighted by an importance factor with given observation Z_t.

For each ith sample target distribution (3.25) is computed as:

$$\alpha\, p(Z_t \mid X_t^{(i)}) p(X_t^{(i)} \mid X_{t-1}^{(i)}, U_{t-1})\, Bel(X_{t-1}^{(i)}) \tag{3.29}$$

Importance factor for each sample is defined as:

$$w^{(i)} = p(Z_t|X_t^{(i)}) \tag{3.30}$$

And given by equation of the quotient:

$$\frac{\overbrace{\alpha\, p(Z_t \mid X_t^{(i)}) p(X_t^{(i)} \mid X_{t-1}^{(i)}, U_{t-1})\, Bel(X_{t-1}^{(i)})}^{\text{target distribution}}}{\underbrace{p(X_t^{(i)} \mid X_{t-1}^{(i)}, U_{t-1})\, Bel(X_{t-1}^{(i)})}_{\text{proposal distribution}}} = \alpha\, p(Z_t|X_t^{(i)}) \tag{3.31}$$

As α is normalizing constant which is proportional to $w^{(i)}$.

Approximation of $Bel(X_t)$ through important sampling is calculated by repeating M times until importance factor is normalized ($\sum_{i=1}^{M} w^{(i)} = 1$). The produced $X_t^{(i)}$ samples define posterior (3.25) discrete probability distribution.

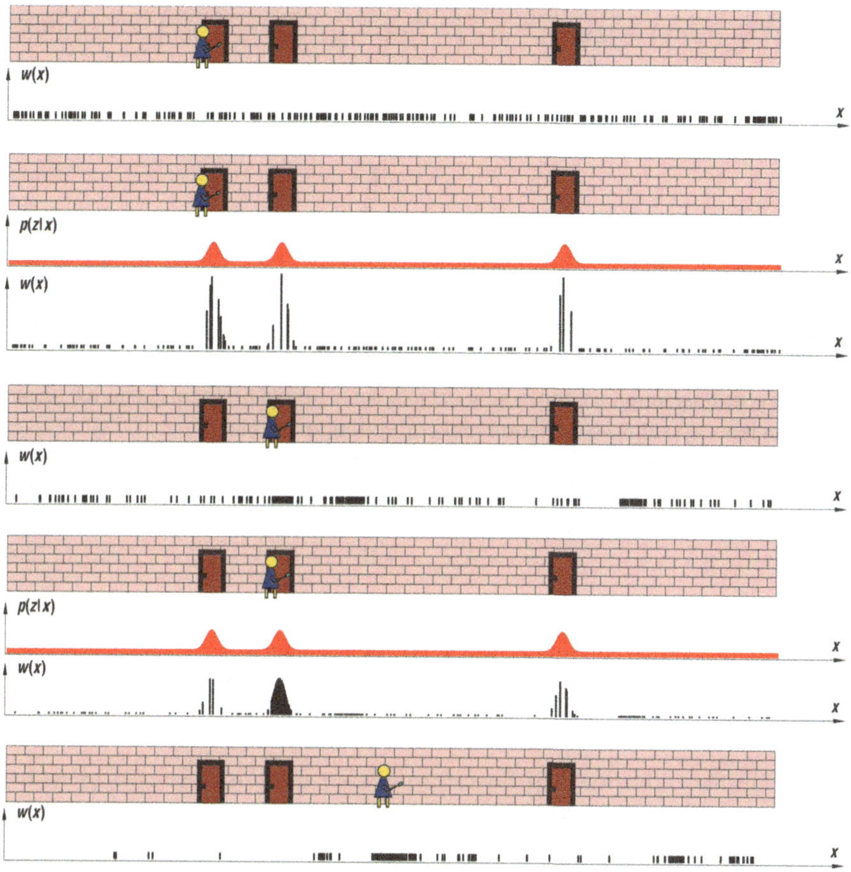

Fig. 3.9 Particle filter localization in one-dimensional space [8]

The simulation of particle filter localization in one-dimension space with three doors environment (similar to shown in Sect. 3.3.1) might be found in Fig. 3.9.

As in the previous example $p(Z_t|X_t)$ that is given from sensor measurements and marked by red, while the probability of $Bel(X_t)$ is black (Fig. 3.9). Here the Belief is presented as discrete distribution of w-weighed samples.

3.3.3 Kalman Filters

The Kalman filters is a group of Bayes filters with state beliefs generally represented as Gaussian distribution. State estimation for all Gaussian filters can be described as:

$$p(X) = |2\pi \boldsymbol{\Sigma}_t|^{-\frac{1}{2}} exp\left(-\frac{1}{2}(X-\mu)^T \boldsymbol{\Sigma}_t^{-1}(X-\mu)\right) \tag{3.32}$$

Equation (3.32) is often use in a short form as:

$$X \sim \mathcal{N}(\mu, \boldsymbol{\Sigma}) \tag{3.33}$$

Where $\mathcal{N}$ stands for *normal distribution*, that is characterized by means of μ with the same number of dimensions as X and covariance matrix Σ dimensioned as $[X \times X]$.

A Gaussian filter developed by Kalman [12]—The *Kalman Filter* unlike the Particular filters based on *linear* state transition and measurements. This means that its base conventional version cannot be used to solve discrete problems. The KF as any Bayesian has two steps—prediction and correction, or update.

Prediction step $\overline{X}_t \sim \mathcal{N}(\overline{\mu}_t, \overline{\boldsymbol{\Sigma}}_t)$ is calculated from previous state μ_{t-1} by the linear state of the transition function $\overline{\mu}_t = f(\mu_{t-1}, U_{t-1})$ that use matrices $\mathbf{A}$ and $\mathbf{B}$:

$$\overline{\mu}_t = \mathbf{A}\mu_{t-1} + \mathbf{B}U_{t-1} \tag{3.34}$$

Covariance matrix $\overline{\boldsymbol{\Sigma}}_t$ is constructed from $\boldsymbol{\Sigma}_{t-1}$ propagated though the linear function with additive noise described as covariance matrix $\mathbf{R}$:

$$\overline{\boldsymbol{\Sigma}}_t = \mathbf{A}\boldsymbol{\Sigma}_{t-1}\mathbf{A}^T + \mathbf{R}_{t-1} \tag{3.35}$$

Next, in the correction step Bayesian posterior mean is calculated by adding predicted mean $\overline{\mu}_t$ to Kalman Gain $\mathbf{K}_t$ multiplied by error of predicted measurements:

$$\mu = \overline{\mu}_t + \mathbf{K}_t(Z_t - \mathbf{C}_t\overline{\mu}_t) \tag{3.36}$$

The linear measurement function $Z_t = h(X_t)$ is defined by zero-mean Gaussian noise with covariance $\mathbf{Q_t}$ and linear sensor model matrix $\mathbf{C_t}$. The Kalman gain is defined as a function of $\boldsymbol{\Sigma}_t$ related to $h(X_t)$:

$$\mathbf{K}_t = \boldsymbol{\Sigma}_t \mathbf{C_t}^T \mathbf{Q_t}^{-1} \tag{3.37}$$

Transformation to form with $\overline{\boldsymbol{\Sigma}}_t$ gives more advantaged Kalman Gain calculation:

$$\mathbf{K}_t = \overline{\boldsymbol{\Sigma}}_t \mathbf{C}_t^T (\mathbf{C}_t \overline{\boldsymbol{\Sigma}}_t \mathbf{C}_t^T + \mathbf{Q}_t)^{-1} \tag{3.38}$$

When all previous calculations have been completed, the final part is to update the posterior covariance:

$$\boldsymbol{\Sigma}_t = \left(\mathbf{C}_t \mathbf{Q}_t^{-1} \mathbf{C}_t + \overline{\boldsymbol{\Sigma}}_t\right)^{-1} = (\mathbf{I} - \mathbf{K}_t \mathbf{C}_t)\overline{\boldsymbol{\Sigma}}_t \tag{3.39}$$

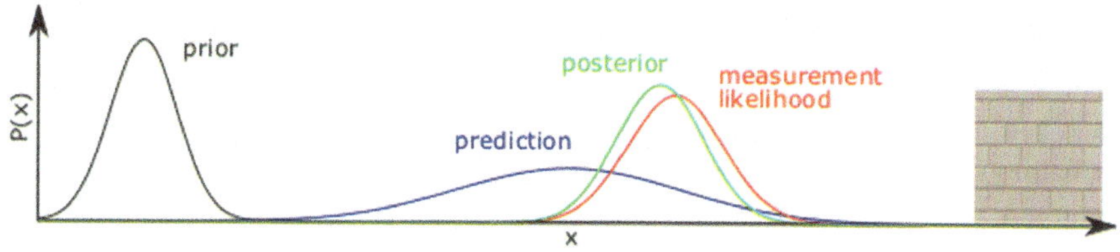

Fig. 3.10 Kalman filter estimation of robot's localization in one-dimensional space [13]

So, for the Kalman filter prediction, belief distribution Eq. (3.26) will have the following form:

$$\overline{Bel}(X_t) = \begin{cases} \overline{\mu}_t = \mathbf{A}\mu_{t-1} + \mathbf{B}U_{t-1} \\ \overline{\mathbf{\Sigma}}_t = \mathbf{A}\mathbf{\Sigma}_{t-1}\mathbf{A}^T + \mathbf{R}_{t-1} \end{cases} \tag{3.40}$$

And posterior belief Eq. (3.27) the following one:

$$Bel(X_t) = \begin{cases} \mu = \overline{\mu}_t + \mathbf{K}_t(Z_t - \mathbf{C}_t\overline{\mu}_t) \\ \mathbf{\Sigma}_t = (\mathbf{I} - \mathbf{K}_t\mathbf{C}_t)\overline{\mathbf{\Sigma}}_t \end{cases} \tag{3.41}$$

Graphical example of simple KF localization in one-dimensional space can be found in Fig. 3.10.

The robot starts the movement from the initial (prior) state in the direction of the wall. A prediction state has little uncertainty because of the only state transition and controls data. When sensors measure the distance to the wall, it is described as measurement likelihood, and is being used to calculate the posterior. Sensor errors are much smaller than in the prior state of transition and control, thus, measurement likelihood uncertainty is lower and posterior is closer to the measurements.

As it has already been mentioned, a regular Kalman filter is linear, so there are also many Kalman Filter extension algorithms that might be used for the *discrete* problems like the *Perturbation Kalman Filter* [14, 15], the *Extended Kalman Filter* [14–16] or *Iterated Extended Kalman Filter* [14]. They all use *linearization* process for state estimation in nonlinear systems. In Sect. 3.5 we shall discuss the EKF as one of most often used filters for solving localization systems in SLAM process.

3.4 Map Estimation

In this section we will focus on most interesting aspect of the book the estimation of map represented as occupancy grid, which basics have been described in previous Chap. 2 (Sect. 2.3.2).

3.4.1 Occupancy Grid

Each cell in the grid is presented by a binary random variable which describes the probability of occupancy by an obstacle—0 of it is free and 1 when is definitely occupied, the unobserved yet cells are fulfilled with some default value, that was being used in a dedicated system (in example 0.5) (Fig. 3.11). Then, the map is given by a product of each cells:

$$p(m) = \prod_{i}^{N} p(m_{x_i,y_i}) \tag{3.42}$$

The calculation of occupation grids map is usually a probabilistic problem of the posterior from the given observations, with pose estimation that is based on the Static State of the Binary Bayes Filter. Therefore, according to Bayes rule and Markov assumption, it is given as:

$$\begin{aligned} p(m_i|Z_{1:t}, X_{1:t}) &= \frac{p(Z_t|m_i, zt-1, X_{1:t})p(m_i|Z_{1:t-1}, X_{1:t})}{p(Z_t|Z_{t-1}, X_{1:t})} \\ &\overset{\text{Markov}}{=} \frac{p(Z_t|m_i, X_t)p(m_i|Z_{1:t-1}, X_{1:t-1})}{p(Z_t|Z_{t-1}, X_{1:t})} \end{aligned} \tag{3.43}$$

After each part of transformations and simplifications that are explained in a detailed way in [1, 17], the final equations will have the form of log odds:

$$p(m_i|Z_{1:t}, X_{1:t}) = 1 - \frac{1}{1 + \exp l_{t,i}} \tag{3.44}$$

$$l_{0:t,i} = l_{0:t-1,i} + \log \frac{p(m_i|Z_{1:t}, X_{1:t})}{1 - p(m_i|Z_{1:t}, X_{1:t})} - \log \frac{p(m_i)}{1 - p(m_i)} \tag{3.45}$$

As we can see, the map accuracy is definitely dependent on the precise pose estimation, just as the successful localization depends on the correct map.

This kind of space representation is usually used for two-dimensional mapping, but is not limited to 2D. There are many techniques that are based on the space mapping by three-dimensional occupancy grids [18–21]. In the Fig. 3.12 we can find two outdoor and indoor 3D occupancy grid maps that are created with the help of the Octomap framework developed by Wurm and Hornung [22, 23].

Fig. 3.11 Cell representation as binary random value

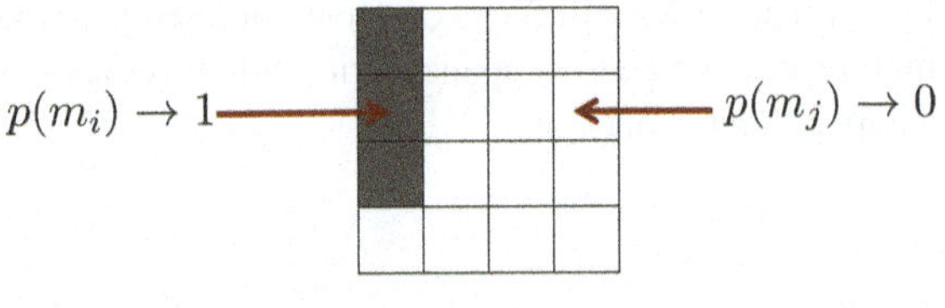

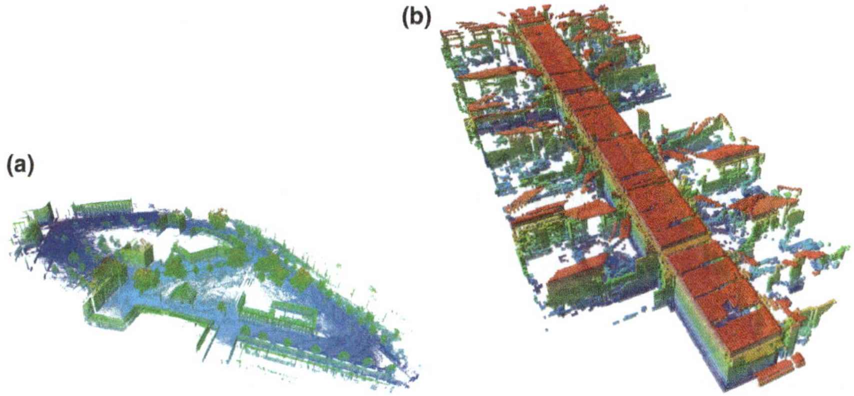

Fig. 3.12 3D occupancy grid maps created with using octomap algorithm and laser rangefinder sensor: **a** shows—the outdoor map of Freiburg University campus (size of the scene: 292 m × 167 m × 28 m), and **b**—the indoor map captured in corridor of Freiburg University building 079 (size of the scene: 43.8 m × 18.2 m × 3.3 m) [23]

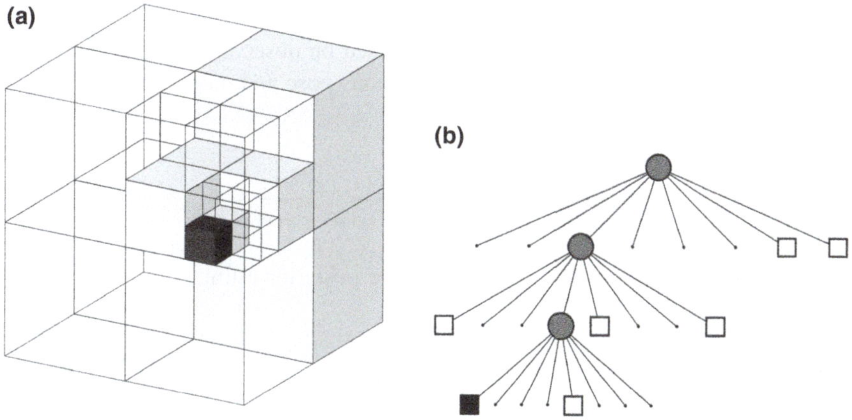

Fig. 3.13 Space compressed in octree volumes with different occupation: **a** volumetric 3D model, **b** its corresponding tree [23]

The presented maps are based on a cloud of points that are gathered by a laser range finder. However, stereo cameras, RGB-D cameras and 3D sonars are also widely used in 3D mapping. Since each of the presented maps consists of a huge number of cells (from many thousands up to billions) for efficient handling of these point sets, the data structure compression is usually used. One of the most efficient representation methods is used in a Wurm's and Hornung's research, called as Octree hierarchical data structure. This algorithm may be seen as continuation of quadtree representation discussed in previous chapter (Sect. 2.3.2). According to this approach, each volume is named as *node*, and has eight child connections with the inner nodes (Fig. 3.13).

Similar basics are used in the applications of the current work. The grids mapping and localization are rich in research approaches for many robotics, both the indoor and the outdoor, path planning solutions.

3.5 Simultaneous Localization and Mapping (SLAM)

3.5.1 SLAM Problem Definition

After discussing the Bayes and Markov assumptions applied to recursive state estimation and probability mapping approaches, is it possible to claim that they are in a close relation and dependence on each other. So, it confirms what has already been mentioned in this book introduction, confirming that for the purpose of efficient mapping and localization, these two calculations must be prepared simultaneously. In this section, the Simultaneous Localization and Mapping, or just SLAM problem definition and its known solutions shall be discussed.

If we put the observed landmarks into a global localization frame—map m, which together with pose also wants to be found. Based on observations and controls for a time from 0 to t, the posterior for simultaneous pose and map estimation can be described by the generic form:

$$p(\underbrace{X_{0:t}}_{\text{robot's path}}, \underbrace{m}_{\text{map}} \mid \underbrace{Z_{1:t}}_{\text{observations}}, \underbrace{U_{1:t}}_{\text{controls}}) \tag{3.46}$$

Applying Bayes rule into (3.46), the recursive posterior estimation might look as follows:

$$p(X_{0:t}, m|Z_{1:t}, U_{1:t}) = \alpha \cdot p(Z_t|X_{0:t}, m, Z_{1:t-1}, U_{1:t}) \cdot p(X_{0:t}, m|Z_{1:t-1}, U_{1:t}) \tag{3.47}$$

Now, if willing to use Markov assumption for simplifying (3.47) into a more clever form, motion and observation models might be easily found:

$$p(X_{0:t}, m|Z_{1:t}, U_{1:t}) = \alpha \cdot \underbrace{p(Z_t|X_t, m)}_{\text{observation model}} \cdot \underbrace{p(X_t|X_{t-1}, U_t)}_{\text{motion model}} \cdot p(X_{0:t-1}, m|Z_{1:t-1}, U_{1:t-1}) \tag{3.48}$$

Let's look into the Eq. (3.48) derivation in more details. The first used assumption is that X_t is a result of all history information from 0 to t that we can encode. It means that probability of observation at time t can be simplified (3.49). Next, the full path splitting into most recent X_t and all previous states $X_{0:t-1}$ by using the same assumption for define simplified prior is wanted (3.50). Derivation of explained simplification is presented below:

$$p(Z_t|X_{0:t}, m, Z_{1:t-1}, U_{1:t}) \overset{\text{Markov}}{=} p(Z_t|X_t, m) \tag{3.49}$$

$$\begin{aligned} p(X_{0:t}, m|Z_{1:t-1}, U_{1:t}) &= p(X_t|X_{0:t-1}, m, Z_{1:t-1}, U_{1:t-1})p(X_{0:t-1}, m|Z_{1:t-1}, U_{1:t-1}) \\ &\overset{\text{Markov}}{=} p(X_t|X_{t-1}, U_t)p(X_{0:t-1}, m|Z_{1:t-1}, U_{1:t-1}) \end{aligned} \tag{3.50}$$

The definition of recursive state and map estimation by the Eq. (3.48) is called a *Full-SLAM* or a *smoothing* problem and its Bayesian network is shown in Fig. 3.14. The estimation here is going through a full path with recovering all possible poses, observations and controls for each time increment.

We could also easily discover that Eq. (3.48) depends on two probability functions related to the already known from this chapter (Sect. 3.3)—the *motion model* and the *observation model*.

This approach is much more profitable when very accurate full path is searched and task has enough computing resources and time for calculations. During previous years many researchers in SLAM community were focusing on robust optimization of smoothing, and its variant *incremental smoothing* ($p(X_{0:t}, m|Z_{1:t}, U_{1:t-1})$), optimization mostly based on the least square problem solutions [13, 21, 24, 25].

The second separate existing SLAM technique is formulation based on considering the most recent pose, the one called *Online-SLAM* or *filtering*:

$$p(X_t, m\,|Z_{1:t}, U_{1:t}) = \int_{X_0} \cdots \int_{X_{t-1}} p(X_{0:t}, m|Z_{1:t}, U_{1:t})dX_{t-1} \cdots dX_0 \tag{3.51}$$

Its Bayesian network graphical representation might be shown as in Fig. 3.15:

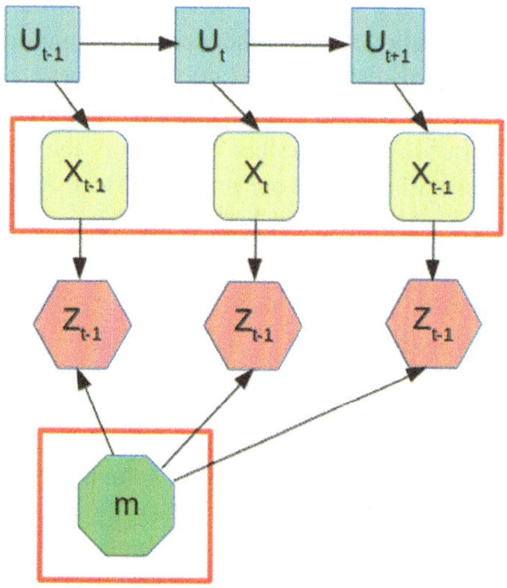

Fig. 3.14 Graphical representation by Bayesian network of full-SLAM

Solution of (3.51) is usually given by recursive integration, one at once:

$$p(X_{t+1}, m\,|Z_{1:t+1}, U_{1:t+1}) = \int_{X_0} \cdots \int_{X_t} p(X_{0:t+1}, m|Z_{1:t+1}, U_{1:t+1}) dX_t \cdots dX_0 \quad (3.52)$$

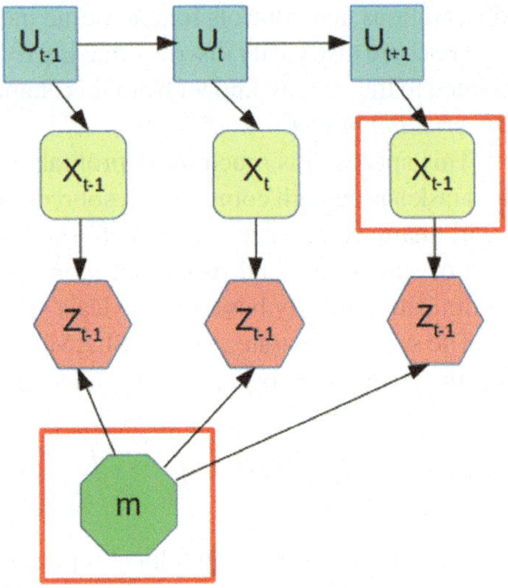

Fig. 3.15 Graphical representation by Bayesian network of online-SLAM

Filtering is a most often used SLAM approach for last two decades, particularly since it allows to have the recent information in a quick time without a huge computing load. This estimation is good for short-term usage and sufficient for path planning or mapping systems that is being applied in autonomous vehicles application usage.

In SLAM researcher community the three main paradigms are existing independently [26]. There are *particle filters*, *Kalman filters* and *graph-based* or *topological* SLAM. We will look at the basic and typical implementations of each SLAM paradigm in next parts of this chapter.

3.5.2 *EKF-SLAM*

The SLAM filtering solution, which is based on the Extended Kalman Filter (EKF) application is the first successfully implemented [27] and most often used Online SLAM algorithm. EKF is similar to original Kalman Filter (described in Sect. 3.3.3), but unlike it, it might be applied in *nonlinear* dynamic systems and for solving discrete problems known like Bayesian recursive state estimation [13–16].

Along with its parent, the state prediction is represented by Gaussian distribution with two-parameters—the mean μ_t and covariance $\overline{\boldsymbol{\Sigma}}_t$. However, for each of the parameters the *nonlinear transition function* $f()$ is being used. The EKF covariance propagation can be done in a similar way to (3.35), but firstly $f()$ must be linearized, usually by the usage of Jacobian matrix $\mathbf{J}$ with derivatives to μ and U. Therefore, the prediction step for EKF will have the following form:

$$\overline{Bel}(X_t) = \begin{cases} \overline{\mu}_t = f(\mu_{t-1}, U_{t-1}) \\ \overline{\boldsymbol{\Sigma}}_t = \mathbf{J}_{\mu_{t-1}} \boldsymbol{\Sigma}_{t-1} \mathbf{J}^T_{\mu_{t-1}} + \mathbf{J}_{U_{t-1}} \mathbf{R}_{t-1} \mathbf{J}^T_{U_{t-1}} \end{cases} \tag{3.53}$$

and the update step use *nonlinear observation function* $h()$ with derivative $\mathbf{H}_t$. In this case, the Kalman gain is represented by:

$$\mathbf{K}_t = \overline{\boldsymbol{\Sigma}}_t \mathbf{H}_t^T (\mathbf{H}_t \overline{\boldsymbol{\Sigma}}_t \mathbf{H}_t^T + \mathbf{Q}_t)^{-1} \tag{3.54}$$

Then, the posterior might be written in the following way:

$$Bel(X_t) = \begin{cases} \mu = \overline{\mu}_t + \mathbf{K}_t (Z_t - h(\overline{\mu}_t)) \\ \boldsymbol{\Sigma}_t = (\mathbf{I} - \mathbf{K}_t \mathbf{H}_t) \overline{\boldsymbol{\Sigma}}_t \end{cases} \tag{3.55}$$

Since SLAM process estimates the map alongside the pose, in EKF-SLAM as the estimated value the *state vector* is used. Its two-dimensional structure consisting of the pose X_t (2.30), the landmarks l_{N_x} and l_{N_y} and signature s_N:

$$Y_t = \{X_t \ m\}^T = \{x \ y \ \theta \ l_{1,x} \ l_{1,y} \ s_1 \cdots l_{N,x} \ l_{N,y} \ s_N\}^T \tag{3.56}$$

Thus, the SLAM posterior Eq. (3.46) has to be described as:

$$p(Y_t | Z_{1:t}, U_{1:t}) \tag{3.57}$$

According to this, mean and covariance of the state vector (3.56) will take the forms:

$$\mu_t = \begin{bmatrix} X_t \\ m \end{bmatrix} \tag{3.58}$$

and

$$\boldsymbol{\Sigma}_t = \begin{bmatrix} \boldsymbol{\Sigma}_{X_t X_t} & \Sigma_{X_t m} \\ \boldsymbol{\Sigma}_{m X_t} & \Sigma_{mm} \end{bmatrix} = \begin{bmatrix} \boldsymbol{\Sigma}_{X_t X_t} & \Sigma_{X_t l_1} & \dots & \boldsymbol{\Sigma}_{X_t l_N} \\ \Sigma_{l_1 X_t} & \boldsymbol{\Sigma}_{l_1 l_1} & \dots & \boldsymbol{\Sigma}_{l_1 l_N} \\ \vdots & \vdots & \ddots & \vdots \\ \boldsymbol{\Sigma}_{l_N X_t} & \Sigma_{l_N l_1} & \dots & \boldsymbol{\Sigma}_{l_N l_N} \end{bmatrix} \tag{3.59}$$

At the initial state in EKF-SLAM they will be filled by zeros, which means that the whole system is in the state of uncertainty. Then, the computing is going through

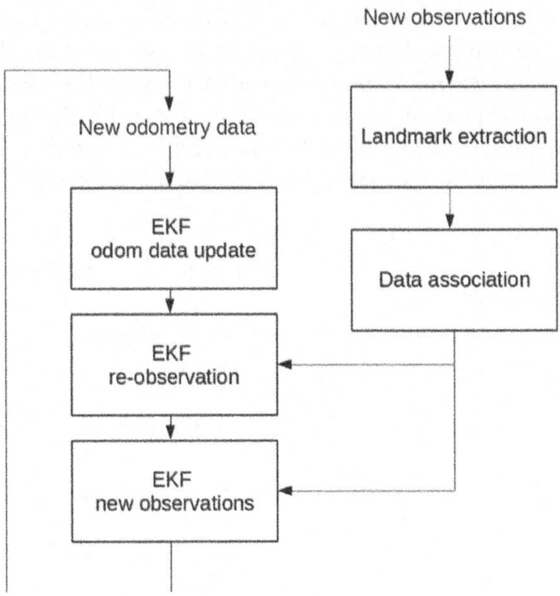

Fig. 3.16 EKF-SLAM algorithm block diagram

three steps similar to regular KF—the *current state estimate* by controls information, the *estimated state update* from observed landmarks, the *new landmarks add to the current state* (Fig. 3.16). It's repeating in a loop with returning μ_t and $\mathbf{\Sigma}_t$ for each calculation time t.

There are many theoretical and practical works dedicated to EKF usage with different approaches. Below a few, as related to the book and previous research experience are mentioned. An EKF-SLAM approach for feature-based maps with extended discussion on EKF-SLAM properties and their practical advantages might be found in the paper [28], the precise odometry technique for a group of mobile robots is presented in [29], the combination of visual or laser odometry with inertial navigation and Kalman filtering is described in [30–32] and for 3D in [33].

3.5.3 Particle Filter SLAM

Another SLAM approach or paradigm is based on *particle filters* usage for *Online-SLAM* problem solution. As has already been discussed in Sect. 3.3.2, the particle filter is based on Bayes rule with Markov assumptions and uses the set of M particles, each ith with importance weight $w^{(i)}$ for drawing the correct pose estimation.

The wide usage of particle filters for SLAM came with a brilliant idea of Rao–Blackwellization method introduction into regular particle algorithm, that was proposed by Doucet et al. in their work [34]. Hence then this approach has been used as base for one of the first efficient particle SLAM implementations known as *Fast-*

SLAM and its later improved—*FastSLAM 2.0*, that was developed by Montemerlo et al. in [35, 36].

Let us look into some details. Since state vector (3.56) is defined by poses and landmarks the key idea is to factorize the SLAM posterior (3.57) into M sampled version, where each particle is a pose with individual landmarks estimated by low-dimensional EKF:

$$p(Y_t|Z_{1:t}, U_{1:t}) = p(X_{0:t}, m_{1:M}|Z_{1:t}, U_{1:t}) = \\ p(X_{0:t}|Z_{1:t}, U_{1:t})p(m_{1:M}|Z_{1:t}, U_{1:t}) = \\ \underbrace{p(X_{0:t}|Z_{1:t}, U_{1:t})}_{\text{particle filter}} \prod_{i=1}^{M} \underbrace{p(m_i|Z_{1:t}, U_{1:t})}_{\text{EKF}} \tag{3.60}$$

Thus, into particle filter solution like MCL [10] into each of the samples, the EKF's belief updated with landmarks is added. The advantages of this approach is in a faster computing time $O(M \log N)$ (EKF – $O(N^2)$), explained by linearization issues excluding and multimodal distribution that can be applied for both grids and landmark-based maps. However, the biggest disadvantage of this method is in losing path history information during resampling step that makes this solution application in systems where state is depended from its history harder.

The Rao–Blackwellized approach in particle filters is represented by a wide spectrum of different implementations like Unscented FastSLAM [37], DP-SLAM, that is recommended by authors for explorations [38], CoreSLAM, which is based on only laser scan mapping, TinySLAM [39]—the shortest SLAM algorithm and optimized for grids maps GMapping [40]. Last algorithm was proposed by Grisetti et al. in [41] and was used as the basis or reference in few of the book related researches [42, 43]. The map created in by GMapping algorithm with additional pose estimation by EKF combined the wheel odometry and IMU localization techniques, as is shown on Fig. 3.17.

3.5.4 Graph-Based SLAM

The graph-based SLAM definition has been first proposed by Lu and Milios in their research publication [44]. This approach is mostly dedicated to Full-SLAM or smoothing problem solution. The SLAM representation in accordance to this paradigm consists of graphs nodes that respond to poses and measurements at each, as they are connected by edges that replace raw measurements (Fig. 3.18).

Edge between nodes X_i and X_j is defined by the spatial constraint error $\mathbf{e_{ij}}$, mean $\mathbf{Z_{ij}}$ and information matrix $\mathbf{\Omega_{ij}}$ related to virtual measurements. The SLAM problem solution is divided into two steps: *graph construction* and *graph optimization* those also called as *back-end* and *front-end* steps (Fig. 3.19).

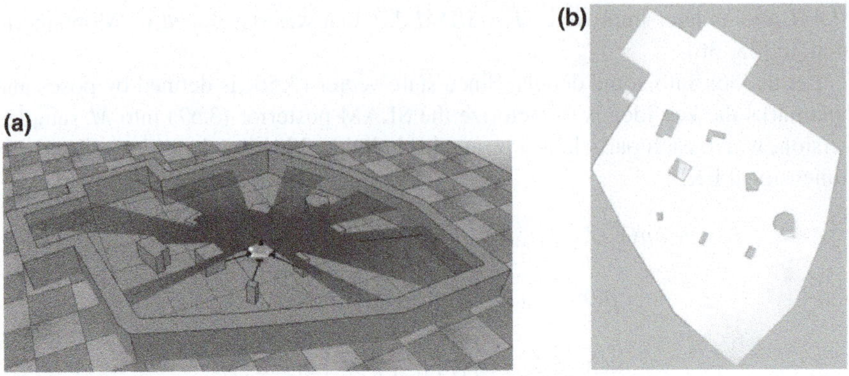

Fig. 3.17 An example SLAM based on particle filter usage (GMapping additionally enhanced by EKF for pose estimation): **a** Test environment simulated in V-REP and **b** its result map calculated by particle filter approach [43]

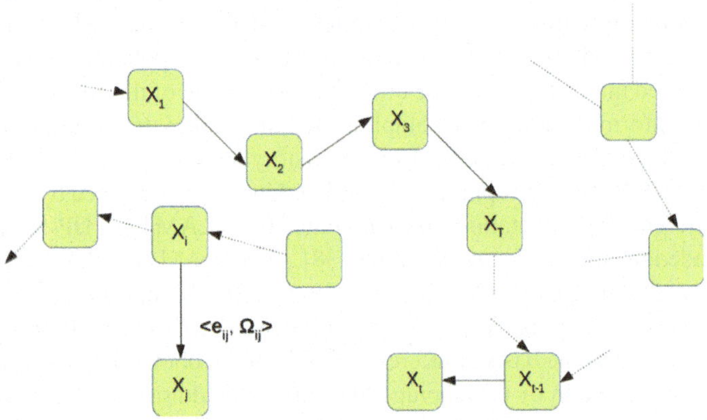

Fig. 3.18 Graph-based representation of SLAM

Fig. 3.19 Graph-based SLAM implementation divided into two back-end and front-end steps: graph construction and optimization

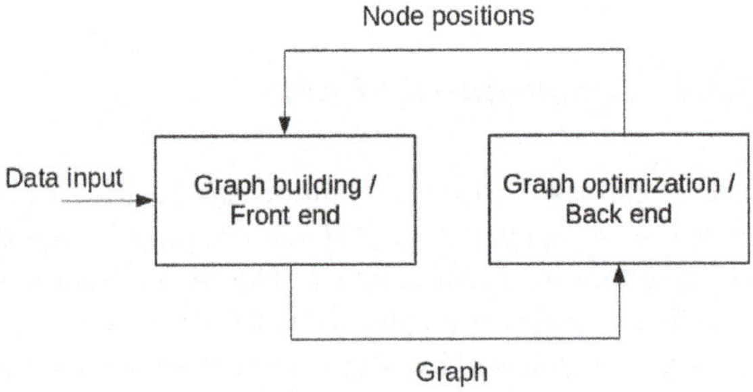

In Gaussian representation (3.32) the *motion* (3.61) and *observation* (3.62) models in Full-SLAM Eq. (3.48) will have the form:

$$p(X_t|X_{t-1}, U_t) = \alpha \exp\left[-\frac{1}{2}(g(X_{t-1}, U_t) - X_t)^T \mathbf{R}_t^{-1}(g(X_{t-1}, U_t) - X_t)\right] \quad (3.61)$$

and

$$p(Z_t|X_t, m) = \alpha \exp\left[-\frac{1}{2}(h(X_t, m) - Z_t)^T \mathbf{Q}_t^{-1}(h(X_t, m) - Z_t)\right] \quad (3.62)$$

Thus, the negative logarithm of Full-SLAM regarding to graph-based SLAM problem might be described as:

$$\begin{gathered} -\log p(X_{0:t}|Z_{1:t}, U_{1:t}) = \\ \text{const} + \tfrac{1}{2}\left[X_0^T \mathbf{\Omega}_0 X_0 + \textstyle\sum_t (h(X_t, m) - Z_t)^T \mathbf{Q}_t^{-1}(h(X_t, m) - Z_t) + \right. \\ \left.(g(X_{t-1}, U_t) - X_t)^T \mathbf{R}_t^{-1}(g(X_{t-1}, U_t) - X_t)\right] \end{gathered} \quad (3.63)$$

As (3.63) can be optimized, the optimization process might be written as:

$$X^* = \underset{\mathrm{X}}{\text{argmin}} - \log p(X_{0:t}|Z_{1:t}, U_{1:t}) \quad (3.64)$$

Let us put it in a more convenient form from the graphs problem point of view. If the prediction of virtual measurements, it is $\overline{\mathbf{Z}}_{\mathbf{ij}}(X_i, X_j)$, then constraint error between observation $\mathbf{Z}_{\mathbf{ij}}$ and $\overline{\mathbf{Z}}_{\mathbf{ij}}$ that can be described in a short form:

$$\mathbf{e}_{\mathbf{ij}}(X_i, X_j) = \mathbf{Z}_{\mathbf{ij}} - \overline{\mathbf{Z}}_{\mathbf{ij}}(X_i, X_j) \quad (3.65)$$

and negative probability logarithm (3.63) of all observations for all pairs in a set $\mathcal{C}$:

$$F(X) = \sum_{\langle i,j\rangle \in \mathcal{C}} \mathbf{e}_{\mathbf{ij}}(X_i, X_j)^T \mathbf{\Omega}_{\mathbf{ij}} \mathbf{e}_{\mathbf{ij}}(X_i, X_j) \quad (3.66)$$

The optimal configuration of nodes X^* is calculated by minimization of (3.66), which is also known as the least square problem solving. The Eq. (3.64) will be:

$$X^* = \underset{\mathrm{X}}{\text{argmin}} \sum_{\langle i,j\rangle \in \mathcal{C}} \mathbf{e}_{\mathbf{ij}}(X_i, X_j)^T \mathbf{\Omega}_{\mathbf{ij}} \mathbf{e}_{\mathbf{ij}}(X_i, X_j) \quad (3.67)$$

or in short form:

$$X^* = \underset{\mathrm{X}}{\text{argmin}}\, F(X) \quad (3.68)$$

The optimization (3.68) is an object of many researches. Optimization is reached by using a wide spectrum of methods for least square problem from standards like the Gauss–Newton and Levenberg–Marquardt [25, 45] to widely used in machine learning Gradient descent methods [46–48].

3.6 Summary

This chapter provided a survey on SLAM basics and latest researches through probabilistic robotics foundations that covers Bayesian filtering introduction, review of precise localization and mapping estimation techniques, discussions on SLAM as a problem itself, and its solution by filtering and smoothing techniques with their approximation methods like least squares problem for Graph-based representation. The focus was mostly on an Extended Kalman and particle filtering represented as Rao–Blackwellized filters paradigms and their applications related multidimensional binary-grids in both indoor and outdoor environment representations and different localization sources.

In accordance to reviewed in previous Chap. 2 localization and navigation methods, it can be concluded that accuracy of mapping and path planning in its big part depends on efficiency of state or pose estimation. The particle filters are precise for pose estimates even in the initial state and deal well with the previous estimation error, but cannot be used for multi-domain data. Kalman Filter suffers from the mentioned advantages of sampling, but can use many information sources and be used with different approaches for estimation increasing. It might be assumed that with Kalman filtering added into pose estimating, it is possible to create a SLAM application system with multiple localization estimation sources, including visual or laser odometry and inertial navigation that might provide more reliable information on the position and orientation in the three dimensions. This assumption opens the discussion on problem analysis of the proposed implementation and its evaluation in Chaps. 4 and 5.

References

1. Thrun, S., Burgard, W., Fox, D.: Probabilistic Robotics. The MIT Press, Cambridge (2005)
2. Plagemann, C., et al.: Gaussian beam processes: a nonparametric Bayesian measurement model for range finders. In: Proceedings of Robotics: Science and Systems (2007)
3. De Laet, T., De Schutter, J., Bruyninckx, H.: Rigorously Bayesian range finder sensor model for dynamic environments. In: Proceedings of the 2008 IEEE International Conference on Robotics and Automation (ICRA), pp. 2994–3001. IEEE (2008)
4. Hähnel, D., Burgard, W., Thrun, S.: Learning compact 3D models of indoor and outdoor environments with a mobile robot. Robot. Auton. Syst. **44**(1), 15–27 (2003)
5. Biber, P., Fleck, S., Straßer, W.: A probabilistic framework for robust and accurate matching of point clouds. In: Proceedings of the Joint Pattern Recognition Symposium, pp. 480–487. Springer (2004)

6. Chen, T., et al.: Likelihood-field-model-based vehicle pose estimation with Velodyne. In: 2015 Proceedings of the IEEE 18th International Conference on Intelligent Transportation Systems (ITSC), pp. 296–302. IEEE (2015)
7. Fox, D., Burgard, W., Thrun, S.: Markov localization for mobile robots in dynamic environments. J. Artif. Intell. Res. **11**, 391–427 (1999)
8. Fox, V., et al.: Bayesian filtering for location estimation. IEEE Pervasive Comput. **2**(3), 24–33 (2003)
9. Thrun, S., et al.: Robust Monte Carlo localization for mobile robots. Artif. Intell. **128**(1–2), 99–141 (2001)
10. Fox, D., et al.: Particle filters for mobile robot localization. In: Doucet, A., de Freitas, N., Gordon, N. (eds.) Sequential Monte Carlo Methods in Practice, pp. 401–428. Springer, Berlin (2001)
11. Kitagawa, G.: Monte Carlo filter and smoother for non-Gaussian nonlinear state space models. J. Comput. Graph. Stat. **5**(1), 1–25 (1996)
12. Kalman, R.E.: A new approach to linear filtering and prediction problems. J. Basic Eng. **82**(1), 35–45 (1960)
13. Sünderhauf, N.: Robust optimization for simultaneous localization and mapping. Ph.D. thesis. Technischen Universitat Chemnitz (2012)
14. Maybeck, P.S.: Stochastic Models, Estimating, and Control. Academic, Cambridge (1979)
15. Grewal, M.S.: Kalman filtering. In: Lovric, M. (ed.) International Encyclopedia of Statistical Science, pp. 705–708. Springer, Berlin (2011)
16. Julier, S.J., Uhlmann, J.K.: New extension of the Kalman filter to nonlinear systems. In: Proceedings of the Signal Processing, Sensor Fusion, and Target Recognition VI, vol. 3068, pp. 182–194. International Society for Optics and Photonics (1997)
17. Milstein, A.: Occupancy grid maps for localization and mapping. In: Jing, X.-J. (ed.) Motion Planning, pp. 382–408. InTech, London (2008)
18. Peasley, B., et al.: Accurate on-line 3D occupancy grids using Manhattan world constraints. In: Proceedings of the 2012 IEEE/RSJ International Conference on Intelligent Robots and Systems (IROS), pp. 5283–5290. IEEE (2012)
19. Johnson, A.E., Kang, S.B.: Registration and integration of textured 3D data. Image Vis. Comput. **17**(2), 135–147 (1999)
20. Badino, H., Franke, U., Pfeiffer, D.: The Stixel world-a compact medium level representation of the 3D-world. In: Proceedings of the Joint Pattern Recognition Symposium, pp. 51–60. Springer (2009)
21. Kohlbrecher, S., et al.: A flexible and scalable SLAM system with full 3D motion estimation. In: 2011 Proceedings of the IEEE International Symposium on Safety, Security, and Rescue Robotics (SSRR), pp. 155–160. IEEE (2011)
22. Hornung, A., et al.: OctoMap: an efficient probabilistic 3D mapping framework based on octrees. Auton. Robot. **34**(3), 189–206 (2013). https://doi.org/10.1007/s10514-012-9321-0, http://octomap.github.com
23. Wurm, K.M., et al.: OctoMap: a probabilistic, flexible, and compact 3D map representation for robotic systems. In: Proceedings of the ICRA 2010 Workshop on Best Practice in 3D Perception and Modeling for Mobile Manipulation, vol. 2 (2010)
24. Guivant, J.E., Nebot, E.M.: Optimization of the simultaneous localization and map-building algorithm for real-time implementation. IEEE Trans. Robot. Autom. **17**(3), 242–257 (2001)
25. Kümmerle, R.: State estimation and optimization for mobile robot navigation. Ph.D. thesis. Universität Freiburg (2013)
26. Stachniss, C.: Robot mapping: introduction to robot mapping (2018). http://ais.informatik.uni-freiburg.de/teaching/ws12/mapping/pdf/slam01-intro.pdf. Accessed 10 May 2018
27. Smith, R.C., Cheeseman, P.: On the representation and estimation of spatial uncertainty. Int. J. Robot. Res. **5**(4), 56–68 (1986)
28. Skrzypczyáski, P.: Simultaneous localization and mapping: a feature-based probabilistic approach. Int. J. Appl. Math. Comput. Sci. **19**(4), 575–588 (2009)

29. Buratowski, T., et al.: The precise odometry navigation for the group of robots. Schedae Informaticae **19**, 99–111 (2015)
30. Williams, B., Reid, I.: On combining visual SLAM and visual odometry. In: Proceedings of the 2010 IEEE International Conference on Robotics and Automation (ICRA), pp. 3494–3500. IEEE (2010)
31. Achtelik, M., et al.: Stereo vision and laser odometry for autonomous helicopters in GPS-denied indoor environments. In: Gerhart, G.R., et al. (ed.) Proceedings of Unmanned Systems Technology XI, vol. 733219-10. International Society for Optical Engineering (2009)
32. Li, M., Mourikis, A.I.: 3-D motion estimation and online temporal calibration for camera-IMU systems. In: Proceedings of the 2013 IEEE International Conference on Robotics and Automation (ICRA), pp. 5709–5716. IEEE (2013)
33. Cole, D.M., Newman, P.M.: Using laser range data for 3D SLAM in outdoor environments. In: Proceedings of the 2006 IEEE International Conference on Robotics and Automation (ICRA), pp. 1556–1563. IEEE (2006)
34. Doucet, A., et al.: Rao-Blackwellised particle filtering for dynamic Bayesian networks. In: Proceedings of the 16th Conference on Uncertainty in Artificial Intelligence. Morgan Kaufmann Publishers Inc., pp. 176–183 (2000)
35. Montemerlo, M., Thrun, S.: FastSLAM 2.0. FastSLAM: a Scalable Method for the Simultaneous Localization and Mapping Problem in Robotics, pp. 63–90. Springer, Berlin (2007)
36. Montemerlo, M., et al.: FastSLAM: a factored solution to the simultaneous localization and mapping problem. In: Proceedings of the AAAI National Conference on Artificial Intelligence. AAAI (2002)
37. Kim, C., Kim, H., Chung, W.K.: Exactly Rao-Blackwellized unscented particle filters for SLAM. In: Proceedings of the 2011 IEEE International Conference on Robotics and Automation (ICRA), pp. 3589–3594. IEEE (2011)
38. Eliazar, A., Parr, R.: DP-SLAM: fast, robust simultaneous localization and mapping without predetermined landmarks. In: Proceedings of the International Joint Conference on Artificial Intelligence (IJCAI), vol. 3, pp. 1135–1142 (2003)
39. Steux, B., El Hamzaoui, O.: tinySLAM: a SLAM algorithm in less than 200 lines C-language program. In: Proceedings of the 11th International Conference on Control Automation Robotics and Vision (ICARCV), pp. 1975–1979. IEEE (2010)
40. OpenSLAM.org. Open SLAM source code community (2018). https://openslam.org/. Accessed 24 Feb 2018
41. Grisetti, G., Stachniss, C., Burgard, W.: Improved techniques for grid mapping with Rao-Blackwellized particle filters. IEEE Trans. Robot. **23**(1), 34–46 (2007)
42. Kudriashov, A., Buratowski, T., Giergiel, M.: Robot's pose estimation in environment exploration process with SLAM and laser techniques. Naukovi Notatki **58**, 204–212 (2017)
43. Buratowski, T., et al.: Robot z laserowym czujnikiem odległości do budowy map 2D. Modelowanie inżynierskie **30**(61), 27–33 (2016)
44. Lu, F., Milios, E.: Globally consistent range scan alignment for environment mapping. Auton. Robot. **4**(4), 333–349 (1997)
45. Grisetti, G., et al.: A tutorial on graph-based SLAM. IEEE Intell. Transp. Syst. Mag. **2**(4), 31–43 (2010)
46. Grisetti, G., et al.: A tree parameterization for efficiently computing maximum likelihood maps using gradient descent. In: Proceedings of the Robotics: Science and Systems, vol. 3, p. 9 (2007)
47. Folkesson, J., Christensen, H.: Graphical SLAM-a self-correcting map. In: Proceedings of the IEEE International Conference on Robotics and Automation (ICRA), vol. 1, pp. 383–390. IEEE (2004)
48. Thrun, S., Montemerlo, M.: The graph SLAM algorithm with applications to large-scale mapping of urban structures. Int. J. Robot. Res. **25**(5–6), 403–429 (2006)

Chapter 4
Multi-agent SLAM-Based Inspection System for Rough Terrain

Abstract This chapter presents the development of an original 2D/3D SLAM application solution for inspection tasks. It covers the fundamentals of mobile robots group structure and behavior including a short survey on formation control in a context of heterogeneous and homogeneous approaches. Multirobot inspection system is seen by Authors as a set of homogeneous robots with decentralized structure and control that cooperate during task perfomance. Then, the integrated approach development for a single robot is presented. Proposed system includes a combined 2D and 3D mapping approach, efficient from both navigation and inspection points of view along with the original robust pose estimation technique, dedicated to rough and isolated terrain, that connects EKF and particle filter techniques. The presented system is taking into account task perfomance in structured and unstructured terrain. The solution includes an original concept and its implementation of work in a multilevel terrain, represented as multi-floral buildings. Next Authors present the development of multirobot inspect system basing on previous proposals, that includes a common map building performed by the image processing techniques. The chapter summary includes an additional short mention of different possible tasks that can be efficiently performed through a developed system implementation.

Keywords Slam · Particle filter · Ekf · 3d Mapping · Robotic inspection · Multi-robot system

4.1 Introduction

This chapter presents a development of an original multi-agent 2D SLAM/3D mapping application solution for exploration tasks. At the beginning, a group of mobile robots fundamentals and path planning concepts are introduced. A short survey on formation control in a context of heterogeneous and homogeneous approaches has been taken. Since the main focus of this book is an integrated SLAM-based mobile robotic approach for a single robot that might be applied to group tasks instead of

A. Kudriashov et al., *SLAM Techniques Application for Mobile Robot in Rough Terrain*, Mechanisms and Machine Science 87, https://doi.org/10.1007/978-3-030-48981-6_4

a multiagent solution development from scratch, the path planning approaches are presented in direct connection to the as far learned basics related to a single robot.

Then, the development of a proposition of SLAM-based, precise pose tracking technique and hybrid 2D/3D mapping system for a single robot can be found in this chapter. The pose estimation technique in accordance to the book Statement is based on laser odometry techniques with a minimal number of used sensors that could reduce power supply and calculation loading and have a positive effect on autonomous operation time. The pose estimation technique is a hybrid solution based on selected techniques, as far presented in this research, that have resistance to known issues and others problems that have been observed during current research. Therefore the calculated pose is presented as a result of pose measurement fusion from different sources. The mapping system includes 2D and 3D world representations, which guarantees safe and robust robot motion in unknown terrain as well as shows workspace characteristics precise enough to be helpful in future work in the already explored terrain. The solution includes an original concept and its implementation of work in a multilevel terrain, represented as multi-floral buildings.

Finally, the proposed integrated solution for a single robot is considered as a control system for homogeneous robots in the group. The analyses of common map building approaches covers centralized and sporadic data exchange during robot meetings—'rendezvous events'. Then, the multirobot SLAM approach, compared to the map merging technique, which sees occupancy grid maps as images as well as a more convenient technique for book narration, has been chosen. The implementation of map exchanging and extending system can be found in this part. The chapter summary includes an additional short mention of different possible tasks that can be efficiently performed through a developed system implementation.

4.2 Group of Mobile Robots

4.2.1 Introduction to Multiagent Systems

Robot's behavior in a group is usually divided by different criteria into *homogeneous* and *heterogeneous*—if they perform the same or different actions [1, 2]. The first heterogeneous robot team was introduced by Fukunda et al. as a cellular robot system (CEBOT), which is able to assemble and disassemble itself [3]. This idea was adopted by many researches, which see the future of mobile robotics in swarm or team approaches [4, 5]. Heterogeneous robotic systems usually require synchronization and often have a restriction of the number of robots in one team due to the roles that they perform [6]. There are different multirobot system architectures, but common approaches could be separated to centralized, hierarchical, decentralized, and hybrid. More details on the homogeneous approach comparison could be found in [6, 7]. An example of unsynchronized team cooperation could be found in [8]. It is easy to find that these approaches may be difficult in implementation and any change of task strategy could take a huge amount of time or even be impossible, which makes the efficiency of heterogeneous group application debatable. The opposite configuration can be met in homogeneous robots—they are usually the same in behavior, equipment, and perform the same roles. Therefore, a configuration that is easy in implementation and potentially not limited by the number of agents [2]. Which group configuration will be better or most efficient regarding the task is always difficult question. Although, heterogeneous robots could be more, or only efficient in many cases, the mentioned requirements and limitations make them not so flexible and easy in development. Therefore, this book assumes the integrated SLAM-based approach as a control system for each agent in a group of homogeneous mobile robots, where all agents in a group have the same configuration and cooperation behavior.

4.2.2 Multirobot Path Planning

Mobile robot navigation basics and selected approaches presented in Chap. 2 (Sect. 2.4) mostly consider motion control by path planning as finding the shortest path between two vertices of graph $\mathcal{G}$, which represents the workspace constructed from $\mathcal{C}$-space, including free spaces and obstacles. This concept is adopted in this book by assuming that workspace decomposition and graph creation is provided by the mapping process. As it was mentioned during both mapping and the navigation survey, cells in occupancy grid maps can be seen as a graph with fixed structure, so Dijkstra's algorithm and its successors might be used for the motion control solution. However, the approaches analyzed so far only assume single robot path search, and applying it by the number of mobile robots in the same workspace is not possible in a direct way. A safe, collision-free path in $\mathcal{G}$ can be delivered by vertex 'reservations', which might be problematic and cause an unbound increase of the full transition time between start and target vertices. Therefore, the robot's motion in a group must be coordinated. Latombe [9] divides multirobot motion planning into two categories: *centralized* and *decoupled*. In the first approach, the $\mathcal{C}$-space of each robot is integrated into one common $\mathcal{C}$-space, where the path search is taken. A centralized probabilistic roadmap solution—the super-graph method for multirobot planning with car-like robots—was introduced by Švestka and Overmars in their works [10, 11]. Peasgood et al. use the spanning tree method for a multiphase planning algorithm that coordinates a group of mobile robots motion in a static world [12]. The survey on multirobot fundamentals can also be found in [13]. The second approach initially calculates individual paths and then tries to resolve potential conflicts, often by coordination with prioritization. The first coordination technique was introduced by Lozano-Perez et al. in [14]. Lozano-Perez's approach found general usage for mobile group control as it allowed for motion via calculated vertices with occasional stops and backward moves for collision avoidance. The significant approach where the A* planner is used for decoupled path planning can be found in [15].

Another group of approaches is based on potential field planning. It applies local minima search in the potential function with fundamentals described in Chap. 2 (Sect. 2.4.5) to the group of mobile robots. The strategy of multirobot motion planning, which is based on a local planner based on potential field planning, was described by Tournassoud in [16]. Kanayama and Hartman, in their work [17], use smooth local path planning to reach the goal. In [18], Wang et al. present single robot path planning as a hybrid approach of global and local planers. The path between the start pose and the goal is calculated by a global planner, which uses static knowledge of the operational world; the dynamic obstacle avoidance is solved by local reaction planning, which is delivered by a potential field planner. This approach, which often is called *hierarchical planning*, connects two different world assumptions—static and dynamic—with their best path planning practice usage—shortest path search and reaction motion planning. Similar solutions can be found in [19–21]. This approach can also be seen as a solution for multirobot collision-free motion, as reactive obstacle avoidance delivers that requirement.

Therefore, the hybrid path planning solution for a multiagent inspection system is proposed in this book. The path between the current localization and the goal might be calculated by the A* algorithm implementation, while the local motion safe behavior could be solved by incremental local planning, where local goals are taken from global plan vertices.

According to knowledge acquired thus far, it can be assumed that the development of a multiagent inspection system, which is based on a group of mobile robots, could be divided into two steps: *single robot system development* and a group of homogeneous robots' integration into *a multiagent system solution*. In the next part of this chapter, these steps will be performed.

4.3 SLAM-Based Robot Inspection Approach

This section covers a development of the SLAM-based integrated solution for an inspection mobile robot, which is equipped with a differential drive motion controller and a laser rangefinder scanner. The system development requires solving several issues related to pose tracking and mapping in different terrains, with focus on rough terrain. Therefore, the development is divided into three steps: *the pose tracking approach*, combined 2D and 3D *mapping techniques* and *integration*. In accordance with the book motivation that could be found in Chap. 1, all steps must be implemented in compatibility with ROS requirements, along with modern software and hardware trends that are based on clean, flexible and scalable systems. The full solution development description starts from the pose tracking approach propose, which can be found in the following subsection.

4.3.1 Pose Tracking

As it was already discussed in previous chapters, all localization methods could be divided into two groups—local and global. Local localization is generally related to the displacement estimation from the robot's motion in a local coordinate frame, which is named in this book as ***'odom'***. The global localization frame is named as ***'map'***. In relations between these frames, drift is usually presented that impacts the precision of fusion of local maps into one global map. Before this error-reducing method will be chosen, local pose estimation has to be discussed.

Regarding to knowledge learned so far in this book, it could be concluded that classical localization methods based on wheel odometry, even with the usage of additional motion sensor information and different estimation techniques, have weak performance in rough or unexpected terrain. Therefore, since the target robotic platform vision sensor system is based on laser range-finders (from the book statement), it is possible to apply the laser odometry incremental localization approach for local pose estimation. The book Statement also assumes that any wheel-origin measurements are a source of additional error and they might be completely removed. Each of the laser odometry approaches analyzed in Chap. 2 (Sect. 2.5.3) is faced with similar problems—drift and data association errors with dynamic changes of Euler angles ϕ and ϑ. In their technique implementations, different authors usually share an interface in the estimator application for additional attached sensors. A well-known and recommended IMU generally causes an error value decrease. However, this is often not enough for a good performance of precise motion in unstructured terrain increment estimation and the global localization problem. In next part of this section, several pose tracking issues related to laser odometry will be surveyed or observed with solution proposes.

One of the problems that occurs in laser odometry approaches is often presented in the motion through long corridors (Fig. 4.1), where the measured obstacles create very similar landmarks to each other and it is hard to determine if the robot is stopped or moving.

It causes huge errors in trajectory estimations by simply making it shorter, which might have an impact on data association during the mapping process (Fig. 4.2). A robot platform equipped with a set of additional sensors that deliver information about displacements could be a solution in some cases. It is possible to use IMU, wheel odometry (only in plane indoor terrain) and GPS (only for outdoor cases). It is easy to find that IMU is a more universal choice. Although, it has worse performance in outdoor terrain and global localization in comparison to GPS, it is independent from the target environment and platform construction.

The other important IMU effect (mentioned in Chap. 2) are precise angle measurements regarding gyroscopes. Orientation errors in laser odometry techniques may be often observed, especially in motion through narrow passages with fast turns (Fig. 4.3).

The laser odometry method, which is suggested for analysis in previous chapters of this book, called the Range Flow-based Approach (RF2O) [22], found significantly better performance in comparison to PL-ICP in the same environments (Fig. 4.4).

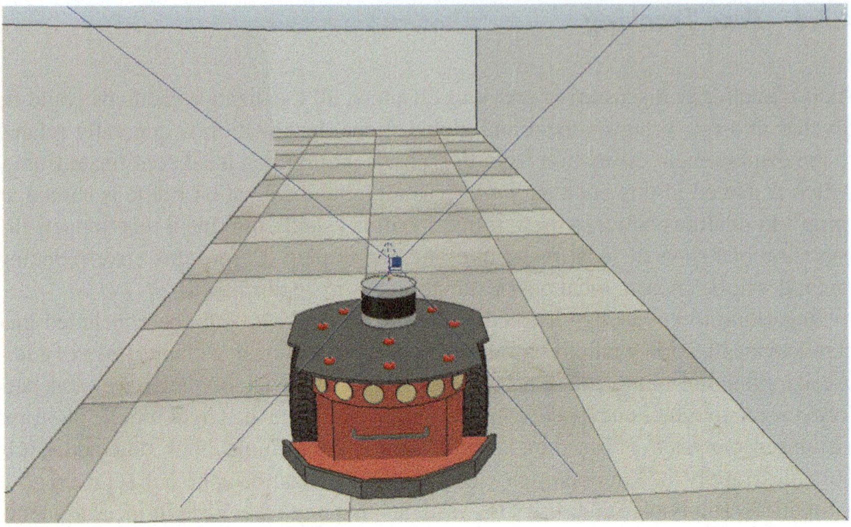

Fig. 4.1 Robot motion in a long corridor: V-REP view

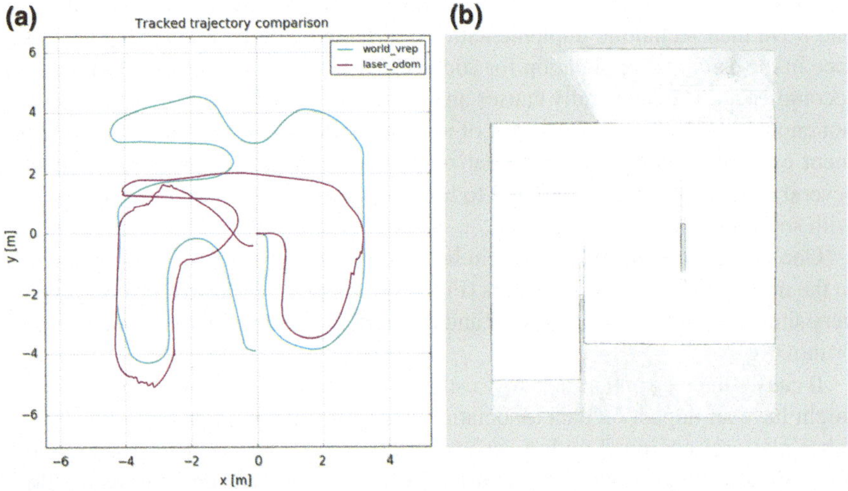

Fig. 4.2 Robot motion in long corridor: trajectory estimated by PL-ICP laser odometry method compared to real passed path (**a**) and resulted map with distortions caused by trajectory errors (**b**)

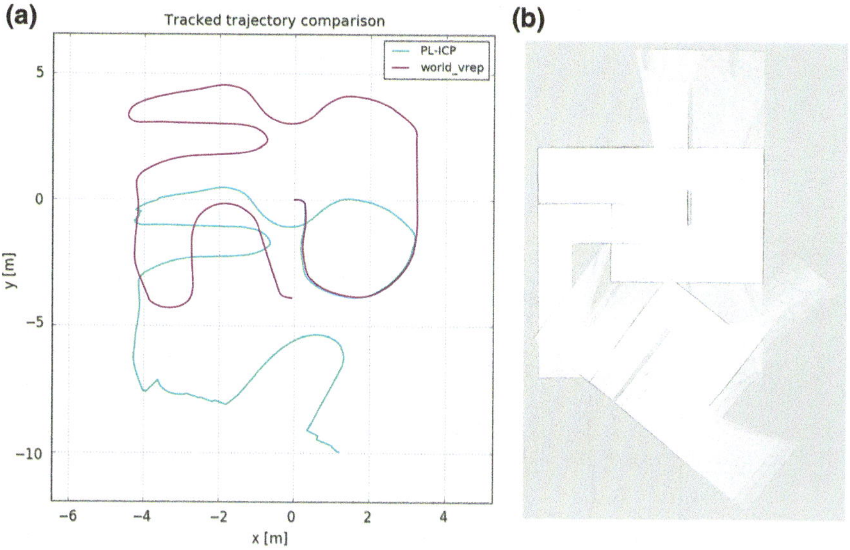

Fig. 4.3 Robot motion in a long corridor: trajectory estimated by the PL-ICP laser odometry method compared to real passed path (**a**) and resulting map with distortions caused by trajectory errors (**b**)

However, these results are still not enough for the book research and a specific estimation technique must be developed. The RF2O approach applies a static world assumption with dynamic obstacle filtering by the Cauchy M-estimator, and the robot's motion is calculated from the motion of observed landmarks provided by the velocity function of sensors. Moreover, it uses one-plane laser scans for $X^{(2)}$ pose estimation, while the book statement demands 3D laser scan measurements. So, applying the RF2O technique in the book research requires an additional mechanism of plane laser measurement conversion from a given point of the cloud. Many 3D scanners have this mechanism implemented natively [23, 24], but because of their costs and for more obstacle detection customization, it needs to be considered as separate process (Fig. 4.5). Plane laser scan data in this process is constructed with relation to the robot's C-space, which means that only points from the cloud that are measured inside of the robot's min. and max. height values are taken for obstacle detection. In this way, if some observed object or obstacle edge is above the highest possible configuration planar path, it will be free, but this object will still be visible for 3D mapping, as original measurements stay untouched.

Since rough terrain mostly assumes unstructured or damaged workspaces (Fig. 4.6), an unexpected dynamic behavior is often present there. It means that in addition to Euler ϕ, and ψ angle dynamic changes, which can be well-tracked by gyroscopes, the robot's base might shift without acting, or its motion might be disturbed by slippage and skidding. In our previous works, the AMCL pose estimation was used for slipping and kidnapping problem solutions where the Kalman Filter fails [25]. Figure 4.7 shows that the usage of the AMCL particle filter could be

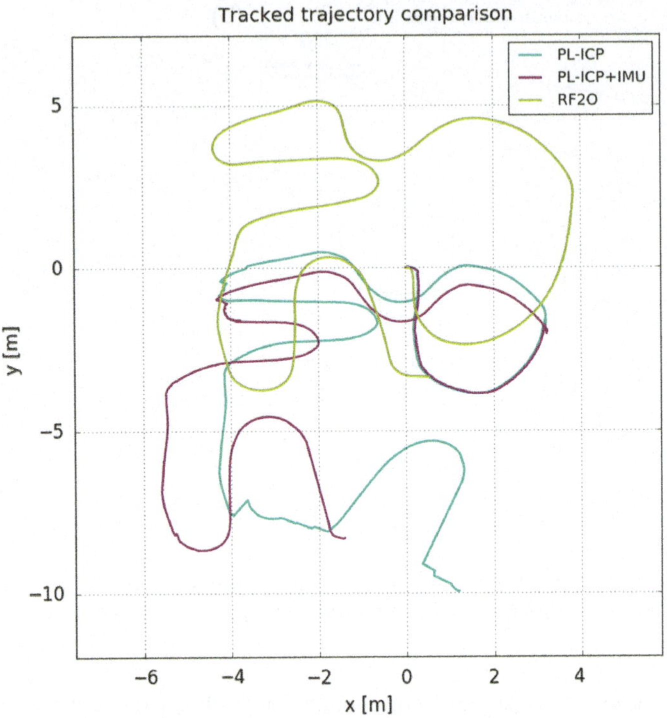

Fig. 4.4 RF2O odometry performance in long and narrow corridors: trajectory estimated by the RF2O laser odometry method compared to the previously used PL-ICP

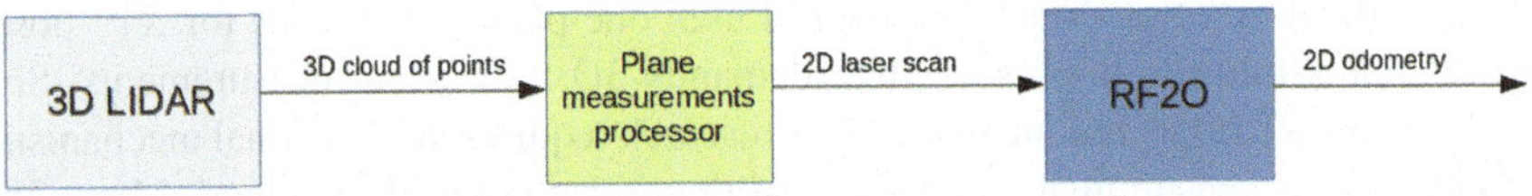

Fig. 4.5 Laser odometry application process measurement flow: the cloud of points from the 3D laser range-finder sensor is converted to plane laser scans by a separate process and then delivered to RF2O

efficient for cases where the estimation process is disturbed by environmental errors like slippage or skidding.

Slippage is a common issue in mobile robotics that causes errors in a whole SLAM process and usually highly impacts on landmark data association. This case is shown in Fig. 4.7b. According to the odometry increments, the robot was dropped outside of its workspace, but when a new observation came, the algorithm redrew a new pose probability and resulting array looks similar to a 'kidnapping' case—the robot was lost for a while and jumped back to the correct localization. This is possible because of the MCL algorithms' nature of pose sampling from known obstacles that gives the ability for efficient global localization, where they are usually

Fig. 4.6 Robot motion in rough terrain: V-REP view

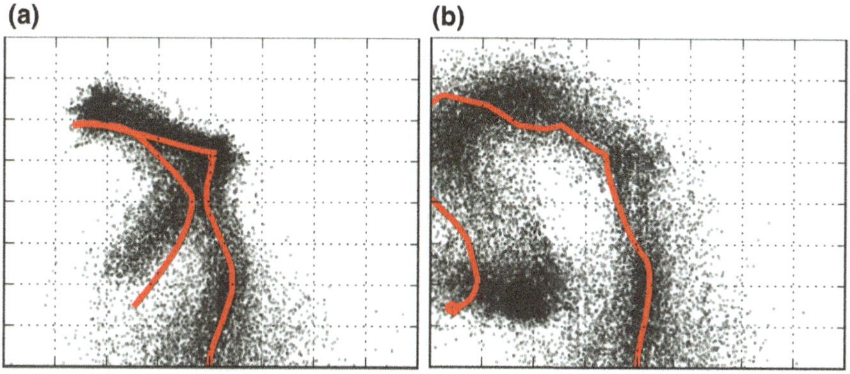

Fig. 4.7 AMCL $X^{(2)}$ pose estimation: wheel odometry (red) and estimated particles (black) in normal motion on a plane surface (**a**) and when the estimation process is disturbed by slippage (**b**)

used. In accordance to this, it is possible to make an assumption that local pose sampling from the observed obstacles by MCL in a local frame could help with cases where odometry and inertial navigation estimated by the Kalman filter usually fail. However, it needs to be mentioned again that Monte-Carlo localization family algorithms cannot use different sources of measurements, so their application only requires odometry information from one source, while Kalman filters can deal with multiple different measurements.

Therefore, it could be decided to implement MCL in the local frame in a following way, where local pose estimation is performed by an Extended Kalman filter, which uses pose measurement inputs from the AMCL particle filter with the given laser odometry data, along with laser scan measurements and incremental techniques of

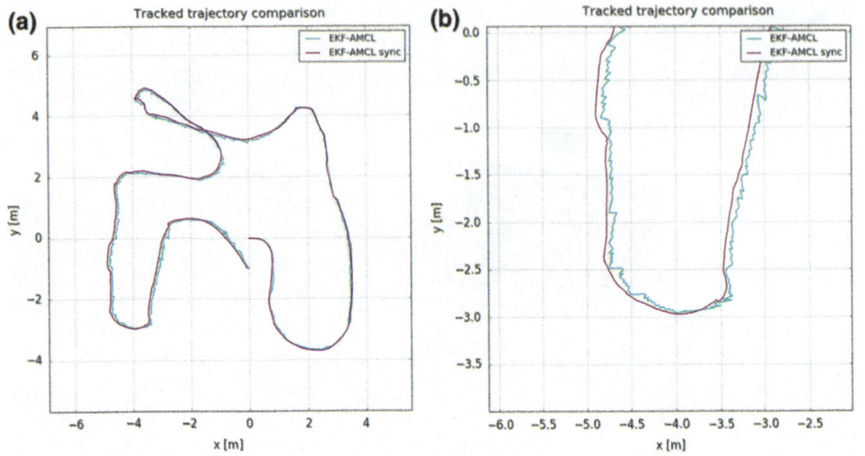

Fig. 4.8 AMCL-EKF trajectory estimation in rough terrain: comparison of synchronized and unsynchronized estimator inputs for a full trajectory (**a**) and its zoom (**b**)

laser odometry and inertial navigation. Since the given measurements are plane, AMCL will estimate the $\mathbf{X}^{(2)}$ pose. A very important thing in this approach is the synchronization of all the measurement, otherwise jump-backs in the estimated value will be present. This is because AMCL calculations are performed after the odometry and after the laser scan data are given, so for the unsynchronized case, some estimates will be based on current and past data (Fig. 4.8).

After the analysis of the existing EKF algorithm implementations, the estimator by Moore and Stouch is chosen as basis for pose estimation in this book [26]. This application implements the open and flexible EKF algorithm with the ROS interface, which in comparison to others [27] offers estimation in 3D space without limitations of the number of sensor inputs and strong requirements for the usage of ROS data message frames.

Local pose tracking will be represented as a product of three different pose sources—laser odometry, IMU and AMCL estimated by EKF, which builds a 3D odometry data structure (Fig. 4.9) that consists of $\mathbf{X}^{(3)}$ pose, coordinate velocities—$\dot{x}$, $\dot{y}$, $\dot{z}$, $\dot{\phi}$, $\dot{\vartheta}$, $\dot{\psi}$ and their covariances. Multi-sensor data fusion is solved by configuration vector customization, where variables are related to coordinates that are calculated from each input. The configuration vector setup for the proposed approach is shown in Table 4.1. The RF2O laser odometry method implementation, beside coordinates on the xy plane, includes tracking for all Euler angles, so these variables might be used in the estimation process. Since, during the experiments, it was found that the RF2O performance for velocity tracking is very weak, the rest of the values were excluded from this input. In accordance with the mentions and conclusions presented as far in this book, all IMU measurement are taken for pose creation. AMCL is represented only by planar x and y coordinates.

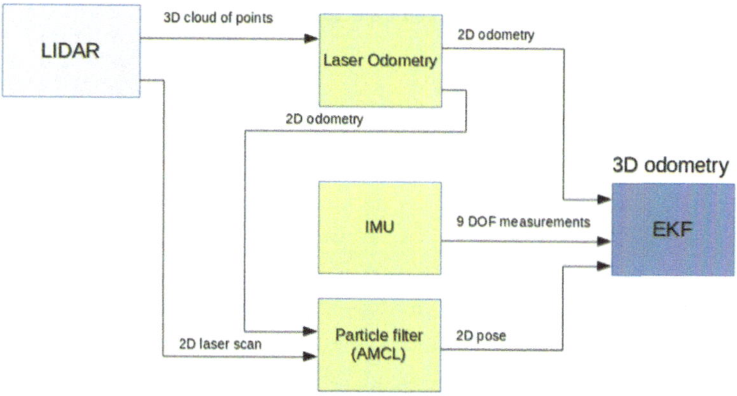

Fig. 4.9 Local pose estimation: combined AMCL-EKF techniques for 3 DOF odometry estimations

Table 4.1 Local pose estimation: extended Kalman filter configuration vector settings

Coordinates tracked

	x	y	z	ϕ	ϑ	ψ	$\dot{x}$	$\dot{y}$	$\dot{z}$	$\dot{\phi}$	$\dot{\vartheta}$	$\dot{\psi}$	$\ddot{x}$	$\ddot{y}$	$\ddot{z}$
RF2O	1	1	0	1	1	1	0	0	0	0	0	0	0	0	0
IMU	1	1	1	1	1	1	1	1	1	1	1	1	1	1	0
AMCL	1	1	0	0	0	0	0	0	0	0	0	0	0	0	0

Since the pose given from the local estimation $\mathbf{X}^{(3)}$ assumes 6 DOF, but regarding kinematic constrains, differential drive robots can operate only in 3 DOF, for path planning and global localization purposes, a state vector might be considered as one

$$\mathbf{X}^{(3)} = \left\{ \mathbf{X}^{(2)} \; z \; \phi \; \vartheta \right\}^T \tag{4.1}$$

which allows to simplify the global localization solution to just the required 3 DOF by $\mathbf{X}^{(2)}$ usage, while z, ϕ and ϑ will be out of the SLAM process and used for other purposes. Global localization in works related to the book [25, 28] is solved by Rao–Blackwellized particle filter SLAM implementation by Grisetti et al. [29]. The legitimacy of this SLAM implementation choice for this book approach is proven by analyses and conclusions from previous chapters. This solution computes recent localization and observations in a space represented as a 2D occupancy grid map. The description of the particle SLAM technique can be found in Chap. 3 (Sect. 3.5.3) and a full description of Grisetti's implementation in a paper [29].

It is very important to empathize that, while in short, different looking scenes, the performance of synchronized and unsynchronized AMCL-EKF estimators differs, but without critical impact on the whole SLAM process and global localization estimation, the impact is presented in huge workspaces with a number of the same

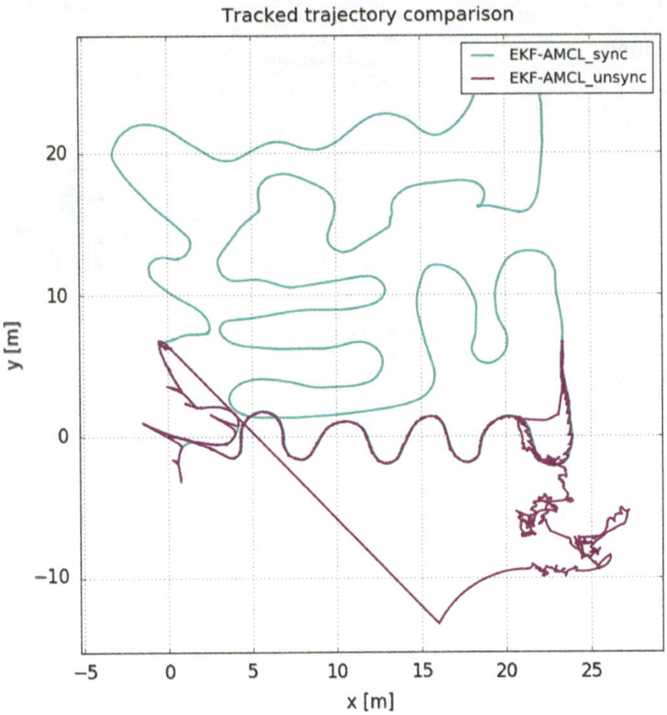

Fig. 4.10 Robot motion in advanced huge terrain: the proposed unsynchronized technique fails and causes a trajectory estimation disaster (EKF-AMCL_unsync), while synchronization improves the estimation accuracy and the whole system safety (EKF-AMCL_sync)

looking landmarks connected to a local 'map' during a long work time (Fig. 4.10). Therefore, it is mandatory to have sensor input synchronization enabled.

To summarize, the robot pose tracking solution proposed in the current book is based on two separate localization techniques related to the local '**odom**' and global '**map**' coordinate frames. The transformation connections between all frames in a mobile robot platform are shown in Fig. 4.11. Where the transformations between the '**map**' and the '**odom**' are performed by global localization Rao–Blackwellized particle filter SLAM, while those between the '**odom**' and the robot's frame '**base**' are calculated by local EKF-AMCL state estimation from the given 3D cloud of points and IMU measurements. All transitions inside of a robot platform construction are static.

After applying the proposed combined estimation technique in the previous cases (Figs. 4.1, 4.6), it is possible to easily conclude that problems presented as far seem to be eliminated on a level, which they do not have enough impact to disturb the whole system (Figs. 4.12 and 4.13). The approach evaluation is described in more detailed in Chap. 5.

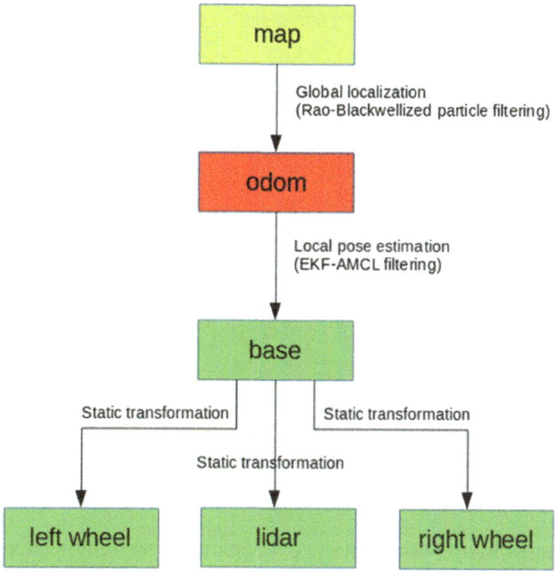

Fig. 4.11 The proposed robot pose estimation solution: connections between coordinate frames and the used transformation techniques

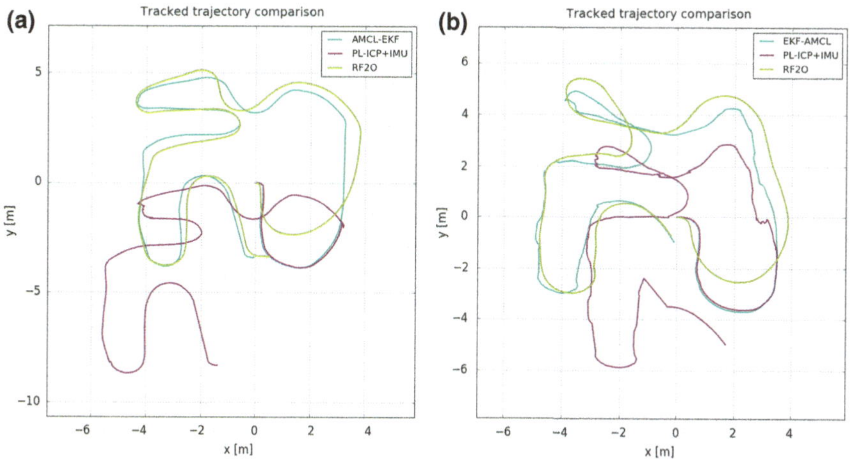

Fig. 4.12 The proposed robot pose estimation solution: an estimated trajectory comparison for a plane terrain with long corridors (**a**) and the same scene with a rough terrain surface (**b**)

4.3.2 *Mapping*

Regarding the book statement, the operating world should be represented by both two- and tree-dimensional occupancy maps. The above-described 2D SLAM process estimates $\mathbf{X}^{(2)}$ pose along with a global 2D grid map consisting of binary cells (Fig. 4.14a). This map can be used for an efficient navigation in both outdoor

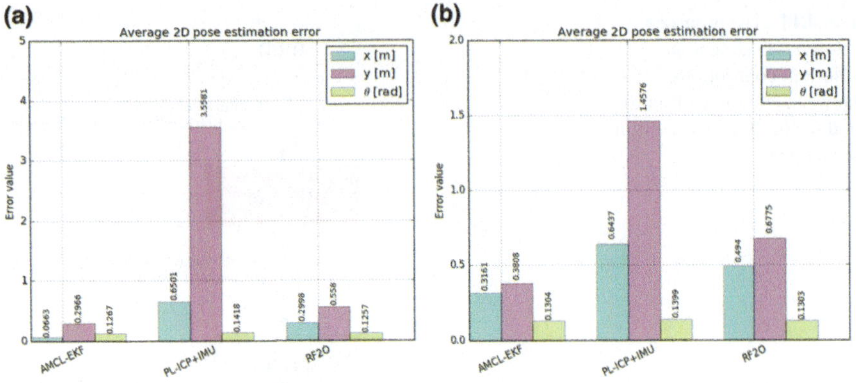

Fig. 4.13 The proposed robot pose estimation solution: a different trajectory estimation technique error comparison for a plane terrain with long corridors (**a**) and the same scene with a rough terrain surface (**b**)

and indoor workspaces, except multilevel environments. For that kind of specific workspace, additional three-dimensional measurements are required. Following the book statement and the prepared survey on 3D occupancy mapping, the octrees approach of the 3D map probabilistic framework implementation by Hornung et al. [30] seems to be a strong candidate. However, it is easy to find that two separate SLAM systems will cause huge ineffective loads on the computing unit and, what is worse, may disturb each other; therefore, dramatically decrease both pose and landmark estimation accuracy. As a consequence, those processes have to be reduced and isolated from each other. During the comparison of two-dimensional maps constructed in the same environment by Octomap SLAM and Rao–Blackwellized particle filter SLAM (Fig. 4.14), it can be seen that the map estimated by the proposed localization and grid mapping solution (Fig. 4.14a) is more accurate than the one estimated by the technique of planar projection from a 3D octomap performed by Hornung's algorithm (Fig. 4.14b).

Therefore, the legitimacy of the previous SLAM system proposition might be consolidated in that way. For the 3D mapping task solution, octomapping can be considered as a half-way 3D SLAM process, performed with already delivered pose without additional pose estimation calculation. An efficient 3D cloud of point data association requires state information that includes all 6 DOF data, since information calculated by AMCL-EKF $\mathbf{X}^{(3)}$ is local and, in the global frame of our system, the only $\mathbf{X}^{(2)}$ that is presented and calculated needs to be merged. The idea is to construct the $\mathbf{X}^{(3)}_{glob}$ value by applying an inverse equation of (4.1) with given global coordinates and z, ϕ, ϑ from the local frame:

$$\mathbf{X}^{(3)}_{glob} = \left\{ \mathbf{X}^{(2)}_{glob} \; z \; \phi \; \vartheta \right\}^T \tag{4.2}$$

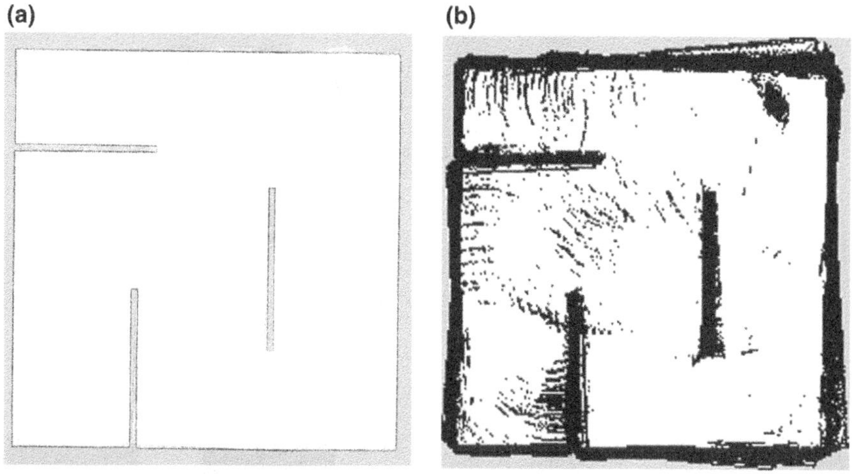

Fig. 4.14 Maps created simultaneously with the proposed combined SLAM technique: Gmapping 2D SLAM (**a**) and octomap 3D SLAM (**b**)

Then, it is possible to use a 3D cloud of points measured by a lidar sensor for the 3D occupancy grid map construction from the known pose. The proposed process block diagram is shown in Fig. 4.15. An octree map estimation may be performed online during exploration or recreated from the collected data as post-processing. The second option is to reduce the loading of computing unit placed in autonomous robot, but this requires data storage equipment with a large enough size for storing laser measurements at each sampling time. In this book, all processes are performed online. The produced scalable octrees map is shown in Fig. 4.16.

As was already mentioned, this approach could be efficiently applied in one-level outdoor or indoor solutions, but it fails in multi-level building inspection tasks, since it generally works in a plane world. This case might be solved with system extension development represented by an additional subsystem for level handling. The possible mechanism idea is based on a dissembled $\mathbf{X}^{(3)}_{glob}$ value, 2D SLAM and 3D cloud of point usage. Three-dimensional workspace planar representation with n-levels can be seen as the sum of each i-level plane map connected by some function $f()$ that describes the geometrical relation between neighboring level 2D maps by the $l_{z,i}$ value, which is related to the distance between level bottom surfaces:

$$\mathbf{M}^{(3)}_{pl} = \sum_{i=1}^{n} f(\mathbf{M}^{(2)}_{i}, \mathbf{M}^{(2)}_{i-1}, l_{z,i}) \tag{4.3}$$

Counting in a chosen notation starts from 0 as level $i = 0$ is related to the ground floor.

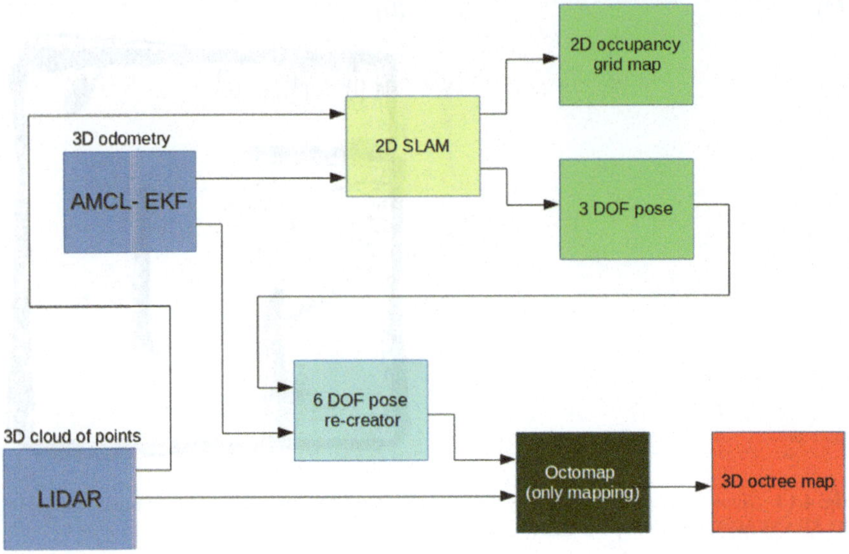

Fig. 4.15 The proposed map building system: 2D and 3D mapping process connection block diagram

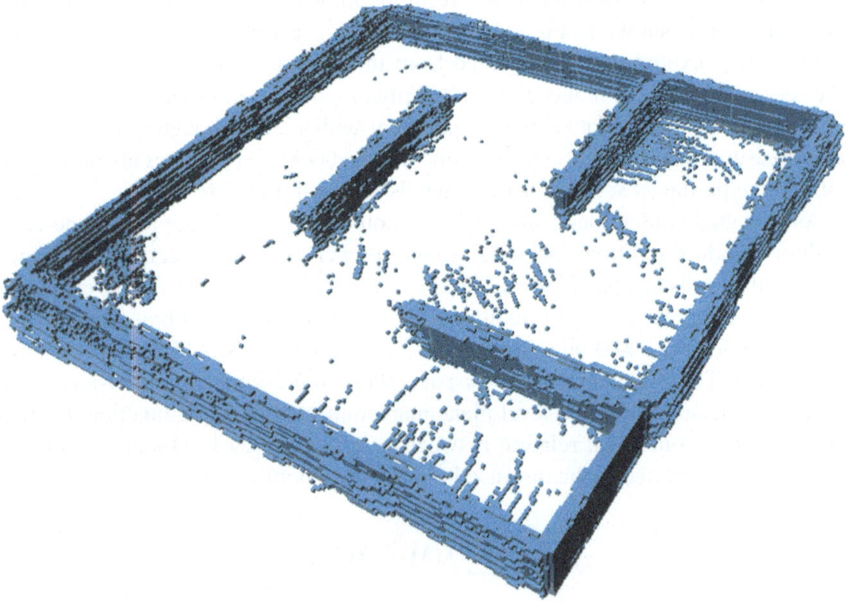

Fig. 4.16 3D occupancy grid map created by proposed technique

The $\mathbf{M}_{pl}^{(3)}$ (Fig. 4.17b), in comparison to 3D space representation by octrees $\mathbf{M}_{ot}^{(3)}$, consists only of projections of each level, while the octomap is fully 3D (Fig. 4.17a). Therefore, multilevel pose tracking (Fig. 4.17c) could be solved in a way of dividing quasi-isolated $\mathbf{X}^{(2)}$ pose estimation in three-dimensional space inside a multi-level buildings (Fig. 4.18) into separate 2D pose and map estimations on each level. The term quasi-isolated is used because, on each level, $\mathbf{X}^{(2)}$ is connected with others by a z coordinate and is a part of the $\mathbf{X}_{glob}^{(3)}$ pose creation. So, the (4.2) tracking in a multi-level environment can be described in a more convenient form as

$$\mathbf{X}_{glob\,0:t}^{(3)} = \sum_{i=0}^{N} \left\{ \mathbf{X}_{glob,i}^{(2)} \; z \; \phi \; \vartheta \right\}_{0:t}^{T} \tag{4.4}$$

where i is related to the level number from 0 to N. Recent $\mathbf{X}_{glob_t}^{(3)}$ value depends on the current level of knowledge.

The switching between levels might be implemented by a '*level database*' application, which stores n number of level information as an L data structure:

$$L = \sum_{i=0}^{n} l_i \tag{4.5}$$

Where each l_i structure consists of floor's ground l_{gnd} and ceiling l_{ceil} levels, which are given from an octree map:

$$l_i = \left\{ l_{gnd,i} \; l_{ceil,i} \; \mathbf{M}_i^{(2)} \right\} \tag{4.6}$$

At each level, initialization states l_{gnd} and l_{ceil} are calculated and the z variable is tracked to be inside these level borders. When z starts to change (raising or decreasing) together with continuing the same direction change of ϕ, for the system, it means that the level is going to be switched. When z is stabilized and under $l_{ceil,i}$ along with ϕ that is close to 0, level switching is performed. Then, the system saves all data with the level's 2D map, resets the 2D map estimation ($\mathbf{X}_{glob}^{(3)}$ pose is continuously kept!) and the initiation procedure is repeated with an additional check whether the level's map already exists in the database. If yes, it is loaded immediately, and the system works as a map update. The 3D octree map construction works in a separate, continuous thread, with conditional independence from 2D SLAM (Algorithms 4.1 and 4.2).

The resulting maps produced by an extension for multi-level building are shown in Fig. 4.17 and a detailed method evaluation is presented in Chap. 5.

In this way, the single robot inspection system is implemented in accordance to all requirements developed in this book. The process product consists of two parts: 3D space reconstructed by an occupancy grip map with the use of octrees, and a 2D building level map, which might be used as cartography documentation. Since all maps are represented as a grid, cells might be considered as a graph with a fixed topography and path planning with the use of Dijkstra's, and its extension

Algorithm 4.1 Multilevel handler. Main loop

```
Require: 2D SLAM, 3D mapping
  i ← 0
  run 2D SLAM handler
  while system runs do
    l_i ← calculate level (M^(3))
    if l_i ∈ L then
      M^(2) ← load level (l_i)
    else
      M^(2) ← 0
    end if
    w ← 2D SLAM handler {w—is returned value by 2D SLAM handler application which is
    associated with level switching}
    if w = 1 or w = −1 then
      l_i.map ← M^(2)
      i ← i + w
    else
      if w = 0 then
        M^(2) ← 0
        restart 2D SLAM
      end if
    end if
  end while
```

Algorithm 4.2 Multilevel handler. 2D SLAM handler

```
Require: 2D SLAM, multilevel handler main thread
  i ← 0
  while system runs do
    if z → l_ceil and φ → π/2 then
      {all z and φ changes observation must take a while}
      repeat
        w ← 1
        return w
      until z ≥ l_ceil and φ = 0
      w ← 0
      return w
    else
      if z → −∞ and φ → −π/2 then
        repeat
          w ← −1
          return w
        until z ≪ l_gnd and φ = 0
        w ← 0
        return w
      end if
    end if
  end while
```

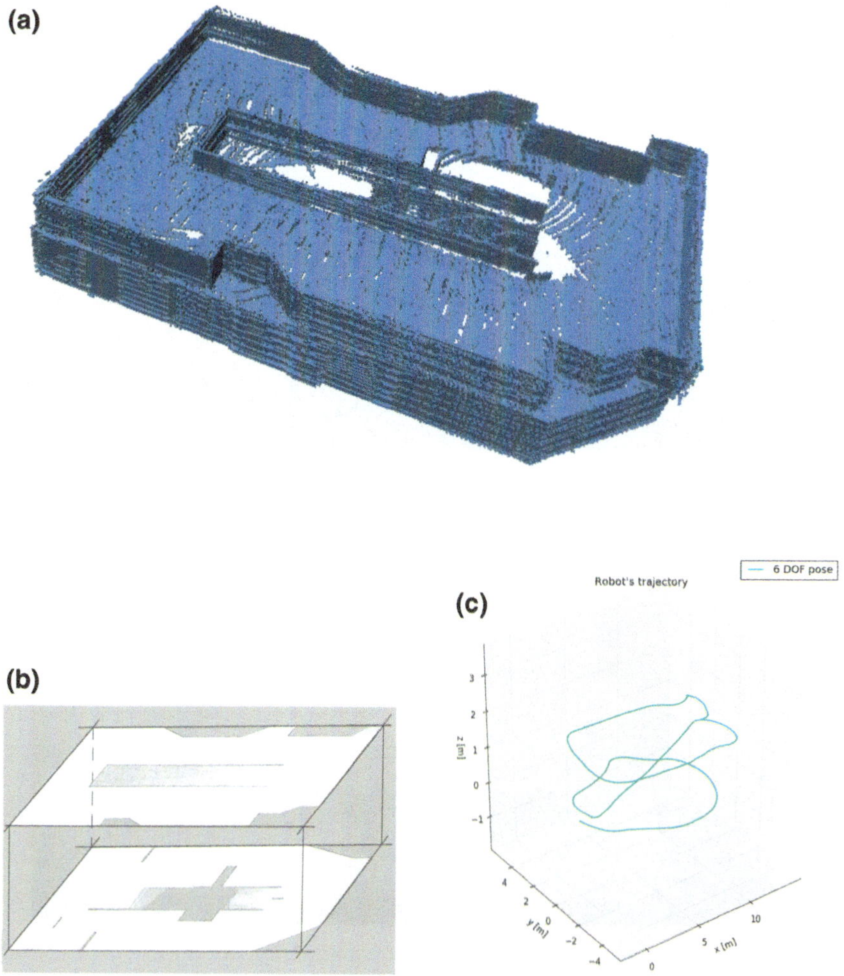

Fig. 4.17 Multi-level exploration system: 3D octomap (**a**), 2D occupancy grid maps of each level with geometric relations (**b**), trajectory passed by robot during exploration (**c**)

algorithms can be easy applied. Basing on all of above, it can concluded that the proposed system and its implementation can be efficiently used for both outdoor and indoor tasks, including multilevel building inspection.

4.4 SLAM for Mobile Robots Group

Developed in the previous section, the SLAM-based inspection approach and its implementation can be seen as a control system for a single robot in a multiagent

(a)

(b)

(c)

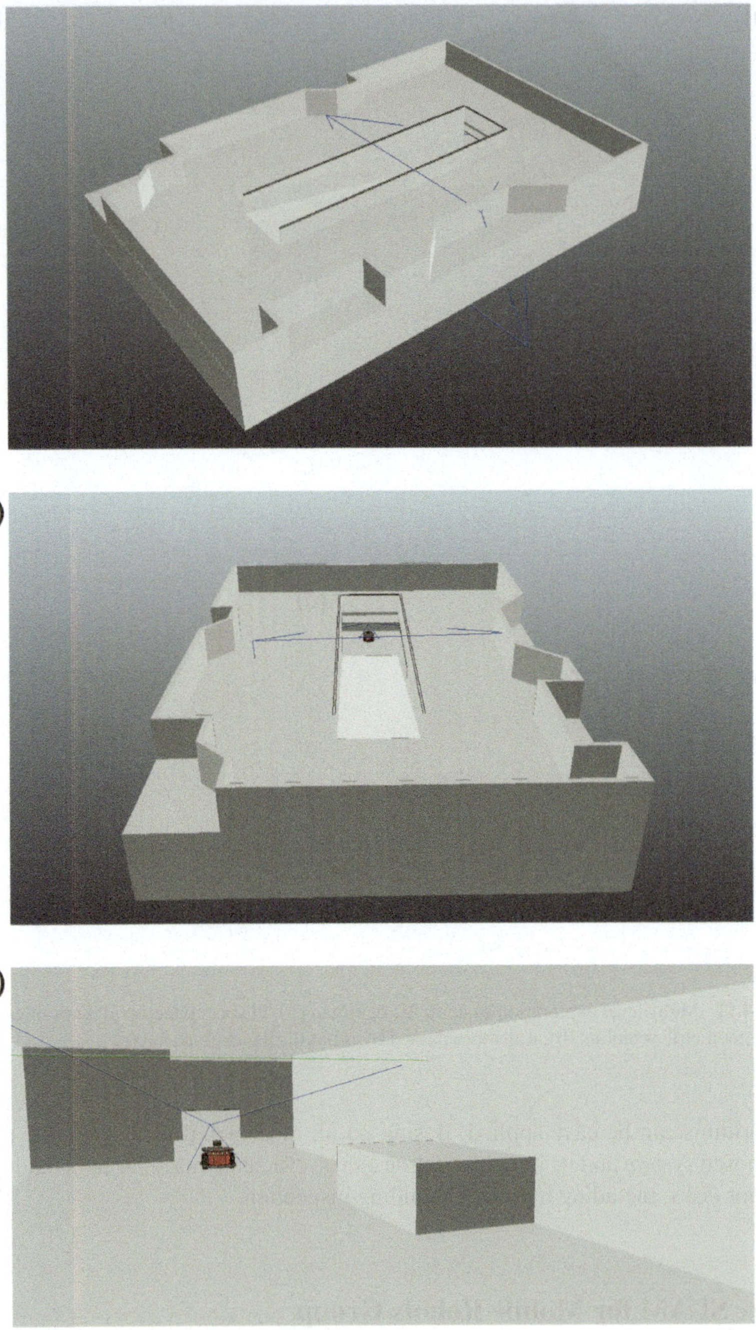

Fig. 4.18 Multi-level building exploration by single robot: VREP scene with differential drive robot and Velodyne VLP-16 LiDAR

inspection system. The multiagent system development could be divided into two separate domains—*multirobot mapping* and *multirobot* navigation. The first one is related to global map creation by a group of mobile robots, and the second is about path planning that is based on global planner handling for efficient exploration and rescue tasks.

4.4.1 Multiagent Localization and Mapping

Multirobot mapping is a relatively new and dynamically developing field in mobile robotics. The main idea can be considered as a problem of global map creation by data association from measurements gathered by a number of mobile robots. Different multirobot SLAM-based approaches are mainly focused on two cases—when robots' initial localization is known or approximated [31, 32] and when it is unknown [33, 34]. The first case is trivial if the robots start from a close localization where they can observe themselves or there is an active beacon global positioning system placed. But in the case when a robot pose is tracked in a robot's global frame (instead of the world's global frame), the problem could be very difficult.

The first notable implementation of a particle filter for multirobot SLAM was in [35] by Howard. It uses the Rao–Blackwellized particle filter SLAM for centralized map processing with each robot pose estimating the common map. It requires a continuous wireless robot communication and huge computing processing, so the number of robot's is strictly limited. When robots meet during exploration (sporadic 'rendezvous event'), a new particle filter instance is triggered in a multirobot global SLAM process. A similar approach has been developed by Fox et al. in [36] with the biggest difference being that robots are looking to meet each other, instead of occasional rendezvous event. An interesting technique of particle filter multirobot SLAM, which improves Howard's algorithm by Rao–Blackwellizing for applications where communication between robots is limited and the initial pose is unknown, is presented in [37] by Carlone et al. In their technique, each robot's SLAM estimation is based on FastSLAM implementation, and a global map is estimated at each rendezvous event by the Rao–Blackwellized particle filter SLAM with each robot's map fuse. When the estimation is completed, the particles restart and robots continue their FastSLAM process from poses before the event. Unfortunately, this technique has a similar number of robot limitations to Howard's technique, and the evaluation in [37] is performed on only two robot group.

Another approach group is based on considering the robot's 2D map as an image. Occupancy grid maps can be seen as images, where cells are associated with pixels; therefore, the map merging problem is seen as image registration, widely known in the computer vision community [38, 39]. Therefore, map merging can be performed through the Hough transform [40]. For given maps $\mathbf{M}_1$ and $\mathbf{M}_2$, the transformation can be found by matrix T with translations tr_x, tr_y and rotation tr_θ:

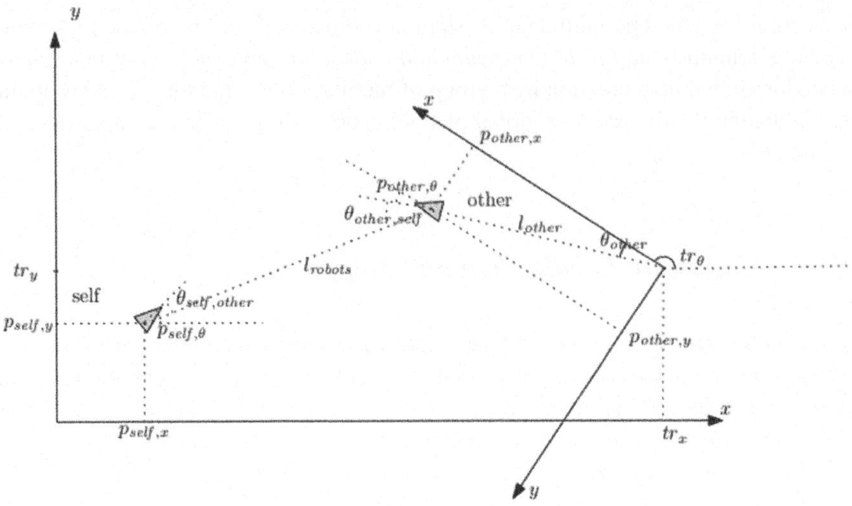

Fig. 4.19 Multirobot map merging using Fast-SLAM approach: robots frame configuration [33]

$$T = \begin{bmatrix} \cos(tr_\psi) & -\sin(tr_\psi) & tr_x \\ \sin(tr_\psi) & \cos(tr_\psi) & tr_y \\ 0 & 0 & 1 \end{bmatrix} \tag{4.7}$$

and described as

$$\mathbf{M}_1 = T \cdot \mathbf{M}_2 \tag{4.8}$$

One of the techniques that uses image processing for map merging is presented by Carpin in [41]. In [33], Özkucur and Akın solve the global mapping problem in a similar way for two robots in a case when they can observe each other and exchange information. Their technique is based on landmark-based metric maps and the FastSLAM technique that use the EKF pose and map estimation.

As received 2D map might be described by each landmark coordinates and covariance (Fig. 4.19):

$$\mathbf{M}_{rec}^{(2)} = \sum_{i=1}^{N} \left\{ l_{x,i}, l_{y,i}, \Sigma_i \right\} \tag{4.9}$$

so each landmark can be transformed by Hugh's transformation matrix (4.7) in a following way:

$$\begin{bmatrix} l_{x,i}{}' \\ l_{y,i}{}' \\ 1 \end{bmatrix} = T \begin{bmatrix} l_{x,i} \\ l_{y,i} \\ 1 \end{bmatrix} \tag{4.10}$$

$$\Sigma i' = T^T \Sigma i$$

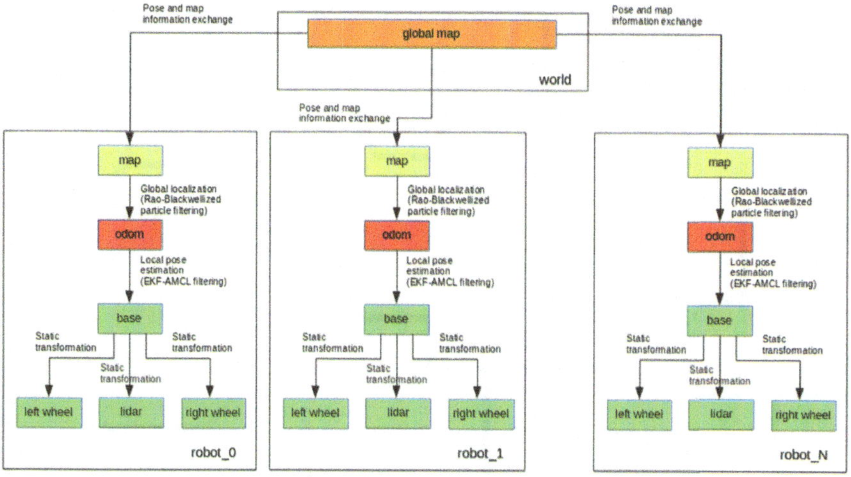

Fig. 4.20 Multirobot mapping: centralized global transformations approach

And then, each landmark is merged into the robot's global map during EKF estimation.

According to the short survey performed above, it could be concluded that all map merging algorithms are challenging in the recreation of the initial pose of each robot. This problem could be solved in various different ways, but one of the most universal and often applied way is data exchange during robot meetings called '*rendezvous*' events. The same basics might be applied to this book's multiagent inspection system. Then, the global world map is considered as a product of each robot's global frame merge (Fig. 4.20). Knowledge on the transformations inside all robot's frame, which is delivered by single robot SLAM-based system developed in the previous Sect. 4.3, allows to reach all measurement data from any to any mobile robot present during this 'meeting'. Moreover, since this system is based on occupancy grid maps, they can be fused by an image processing algorithm via Hough's transformation matrix. Therefore, there is no special need for an additional centralized, global multi-robot SLAM process that covers all robots in the group.

Map merging application can placed at each robot or isolated to centralized system. First approach gives more universal system, but requires additional loading on each robot's calculation units. The second could be solved by extending active beacons to map information (Fig. 4.21). In that case, the data exchange will occur at any time when the robot is close enough to the beacons. However, this solution might be applied only in terrain where active beacons are placed, so it is very expensive and it has no practical solution for workspaces discussed in this book.

The information exchange during sporadic meetings seems to be a more practical solution (Fig. 4.22). As for the map merge algorithm, a simple implementation of Carpin's algorithm [41] with the ROS package wrapper 'multirobot-map-merge' [42] usage has been applied.

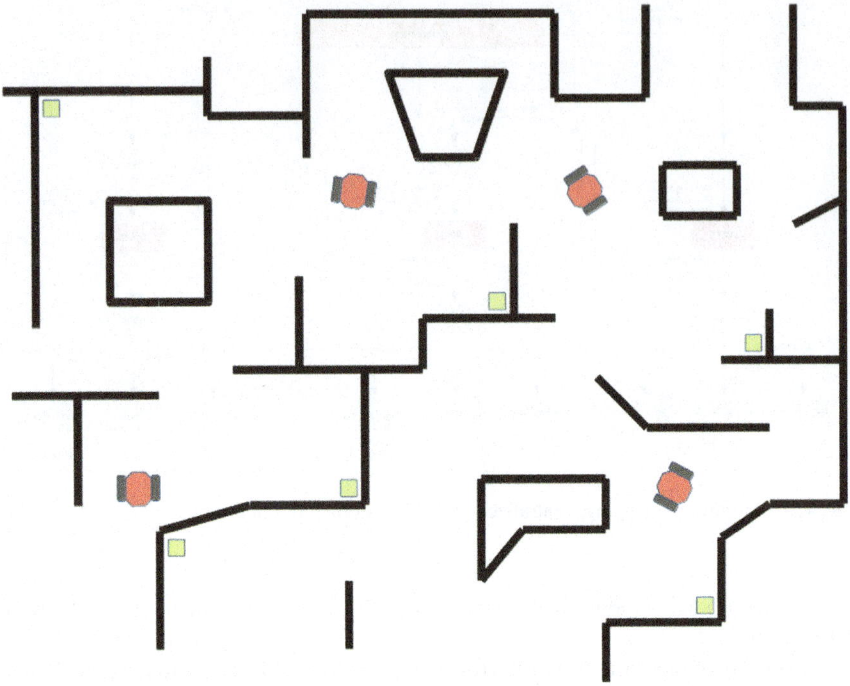

Fig. 4.21 Multirobot mapping merge: active beacons workspace

The multirobot mapping approach then must to be solved by the following algorithm:

Algorithm 4.3 Map merging. Data exchange during 'rendezvous' event

robot ← false
while system runs **do**
  robot ← check if robot in range
  while robot = true **do**
    SEND(exchange request)
    response ← RECEIVE(timeout)
    if response = ready **then**
      SEND($\mathbf{M}^{(2)}$, $\mathbf{X}^{(2)}$)
      $\mathbf{M}_{rec}^{(2)}$, $X_{rec}^{(2)}$ ← RECEIVE(timeout)
      $\mathbf{M}^{(2)}$ ← merge maps ($\mathbf{M}^{(2)}$, $X^{(2)}\ T$, $\mathbf{M}_{rec}^{(2)}\ X_{rec}^{(2)}$)
    else
      SEND(timeout message)
    end if
  end while
end while

Fig. 4.22 Multirobot mapping merge: sporadic 'rendezvous' exchange approach

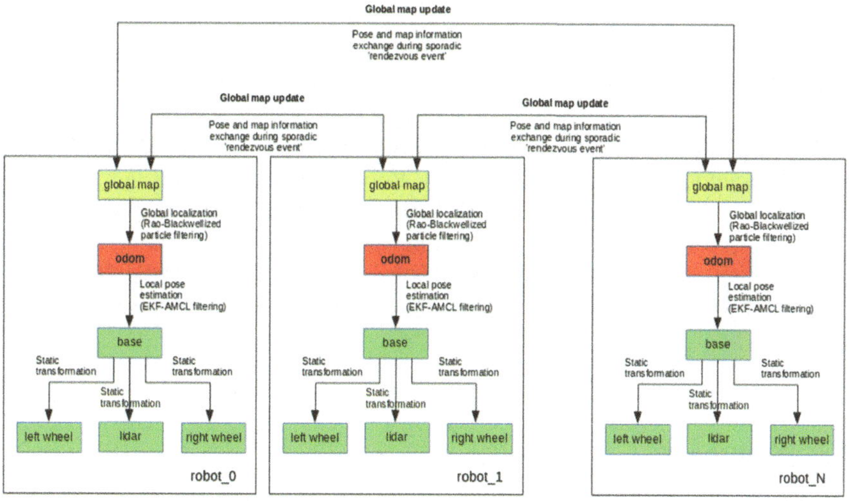

Fig. 4.23 Multirobot mapping: proposed global transformations approach

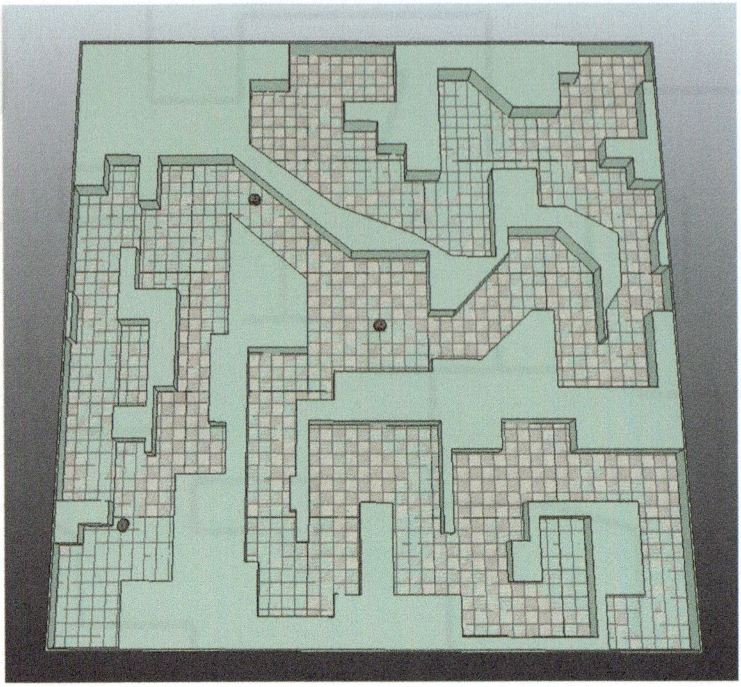

Fig. 4.24 Multirobot mapping: V-REP scene with tree differential drive controlled mobile robots

Thus, in this approach, the robot's global map will be de-facto the world's global map and it will be updated during sporadic 'rendezvous' events, and the number of used robots is theoretically not limited (Fig. 4.23).

The map merging system performance can be demonstrated by the V-REP scene with three autonomous robots (Fig. 4.24). Robots navigated by the A* algorithm started from 3 different corners of scene with a target pose at the center of the scene. The maps of each robot before the meeting are shown in Fig. 4.25a–c. The global common map calculated after the 'rendezvous' event is shown in Fig. 4.25d.

A detailed evaluation of the presented multiagent exploration system can be found in Chap. 5.

4.5 Summary

The integrated SLAM-based solution for mobile robot development presented in this chapter is a working implementation of a multirobot inspection system for rough terrain without a limitation of the number of involved agents. The system development could be divided into separate parts, which represent each field of research performed during the book research. There are a single robot pose estimation, 2D mapping, 3D

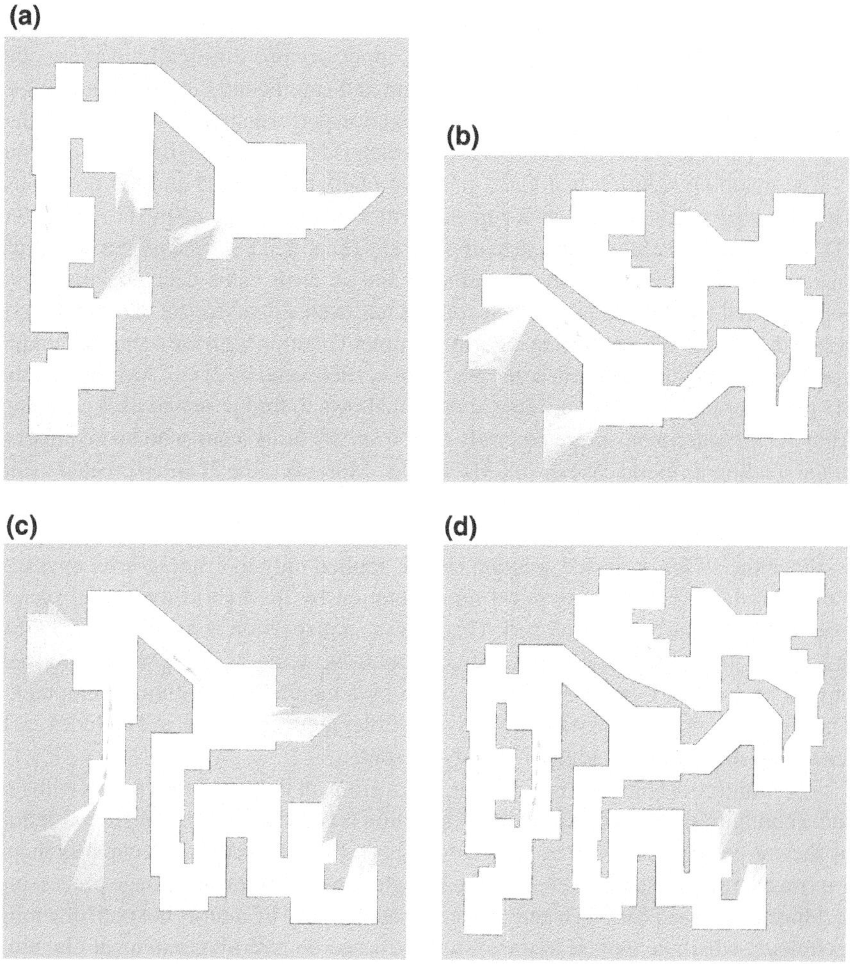

Fig. 4.25 Multirobot mapping: 2D maps created by each of robots (**a**)–(**c**) and resulted merged (**d**)

mapping, multilevel mapping and pose tracking, multirobot cooperation, navigation and map merging.

Basing on the pose estimation and localization technique research performed during this book as well as previous works, the novel local odometry estimation technique that is based on hybrid EKF and AMCL estimators has been proposed. The robot recent pose estimated in the odometry frame is two-dimensional and represented as a product of data fusion, where incremental laser odometry and inertial navigation, used together with the particle filter approach localization estimate, are delivered to the EKF for further estimation. Since the MCL calculation is based on observations from LIDAR measurements and the laser odometry technique, which also use those measurements, the developed technique strongly requires for the input data to be synchronized. The synchronization of input streams drastically increases the precision of pose estimation in a rough terrain and significantly in different cases of plane-structured terrain. For the global localization performance, it has been decided to use the particle-based SLAM approach, which is based on Rao–Blackwellization.

The mapping solution is divided into two quasi-isolated 2D and 3D processes. The 2D map, represented as an occupancy grid, is delivered online by the mentioned 2D SLAM process. Since the usage of different 2D and 3D SLAM processes simultaneously causes an estimation disturbance and an error value decrease, they have been isolated from themselves. That conflict has been solved by the 3D SLAM process reducing to only a mapping algorithm implementation with the octree technique usage like the postprocess where the exact pose is delivered by 2D SLAM through the 3D pose recreation algorithm. That approach allows to find pose and map estimates from 3D measurements more precisely than from 2D laser scans with less computational loading as in the case of full 3D SLAM. Moreover, the 2D map created using this approach is more accurate than a planar projection from 3D SLAM, while the precision of the 3D map is the same.

Since the so far proposed solution can be applied only to a single-level environment, the three-dimensional world representation by the hybrid multilevel planar map approach has been developed. This allows an inspection in a multi-floral building where global 3D pose is tracked continuously, while 2D SLAM is performed only at each level. Level changing is solved by a handling algorithm development. The algorithm is based on robot's z-axis and Euler angles tracking and includes each level map storage and loading if already present.

The multirobot approach is represented as a group of homogeneous robots with the same configuration and equipment. Cooperation is based on a common map creation by the current map and pose exchange during occasional meetings. Occupancy maps are considered as images; therefore, the Hough transformation and image processing techniques are used for given and received map merge. The multirobot path planning technique, which is applied in the system, is based on hybrid hierarchical planning with the A* algorithm as a global planner, and for local motion planning with dynamic obstacle avoidance, the potential field planner algorithm has been applied.

References

1. Brooks, R.A.: Challenges for complete creature architectures. In: Proceedings of the 1st International Conference on Simulation of Adaptive Behavior, pp. 434–443 (1991)
2. Bongard, J.: Divide and Conquer: A Novel Approach to the Design of Heterogeneous Multi-robot Systems. University of Sussex (1999)
3. Fukuda, T., et al.: Structure decision method for self organising robots based on cell structures-CEBOT. In: 1989 Proceedings of the IEEE International Conference on Robotics and Automation (ICRA), pp. 695–700. IEEE (1989)
4. Stroupe, A., et al.: Heterogeneous robotic systems for assembly and servicing. In: Proceedings of the International Symposium on Artificial Intelligence, Robotics and Automation in Space. European Space Agency (2005)

5. Arkin, R.C., Balch, T.: Cooperative multiagent robotic systems. Auton. Robot. **1**(1), 27–53 (1994)
6. Parker, L.E.: Multiple mobile robot systems. In: Siciliano, B., Khatib, O. (eds.) Springer Handbook of Robotics, pp. 921–941. Springer, Berlin (2008)
7. Quinn, M.: A comparison of approaches to the evolution of homogeneous multi-robot teams. In: 2001 Proceedings of the IEEE Congress on Evolutionary Computation, vol. 1, pp. 128–135. IEEE (2001)
8. Grabowski, R., et al.: Heterogeneous teams of modular robots for mapping and exploration. Auton. Robot. **8**(3), 293–308 (2000)
9. Latombe, J.-C.: Robot Motion Planning. Springer Science & Business Media, Germany (2012)
10. Švestka, P., Overmars, M.H.: Coordinated path planning for multiple robots. Robot. Auton. Syst. **23**(3), 283–292 (1998)
11. Švestka, P., Overmars, M.H.: Probabilistic path planning. In: Laumond, J.-P. (ed.) Robot Motion Planning and Control, pp. 255–304. Springer, London (1998)
12. Peasgood, M., Clark, C.M., McPhee, J.: A complete and scalable strategy for coordinating multiple robots within roadmaps. IEEE Trans. Robot. **24**(2), 283–292 (2008)
13. Parker, L.E.: Path planning and motion coordination in multiple mobile robot teams. In: Meyers, R.A. (ed.) Encyclopedia of Complexity and System Science, pp. 5783–5800. Springer, Berlin (2009)
14. O'Donnel, P.A., Lozano-Perez, T.: Deadlock-free and collision-free coordination of two robot manipulators. In: Proceedings of the IEEE International Conference on Robotics and Automation (ICRA), pp. 484–489. IEEE (1989)
15. Bennewitz, M., Burgard, W., Thrun, S.: Optimizing schedules for prioritized path planning of multi-robot systems. In: Proceedings of the IEEE International Conference on Robotics and Automation (ICRA), vol. 1, pp. 271–276. IEEE (2001)
16. Tournassoud, P.: A strategy for obstacle avoidance and its application to multi-robot systems. In: 1986 Proceedings of the IEEE International Conference on Robotics and Automation, vol. 3, pp. 1224–1229. IEEE (1986)
17. Kanayama, Y., Hartman, B.I.: Smooth local path planning for autonomous vehicles. In: Cox, I.J., Wilfong, G.T. (eds.) Autonomous Robot Vehicles, pp. 62–67. Springer, Berlin (1990)
18. Wang, L.C., Yong, L.S., Ang, M.H.: Hybrid of global path planning and local navigation implemented on a mobile robot in indoor environment. In: Proceedings of the IEEE International Symposium on Intelligent Control, pp. 821–826. IEEE (2002)
19. Knepper, R.A., Srinivasa, S.S., Mason, M.T.: Hierarchical planning architectures for mobile manipulation tasks in indoor environments. In: Proceedings of the IEEE International Conference on Robotics and Automation (ICRA), pp. 1985–1990. IEEE (2010)
20. Tarokh, M.: Hybrid intelligent path planning for articulated rovers in rough terrain. Navigation **15**, 17 (2008)
21. Kambhampati, S., Davis, L.: Multiresolution path planning for mobile robots. IEEE J. Robot. Autom. **2**(3), 135–145 (1986)
22. Jaimez, M., Monroy, J., Gonzalez-Jimenez, J.: Planar odometry from a radial laser scanner. A range flow-based approach. In: Proceedings of the IEEE International Conference on Robotics and Automation (ICRA), Stockholm, Sweden, pp. 4479–4485. IEEE (2016)
23. Velodyne Inc. VLP-16 Manual: User's manual and programming guide (2016). http://www2.velodynelidar.com/l/208822/2018-02-23/6t1jp?File_Name=ManualVLP16&File_Code=ManualVLP16. Accessed 15 June 2018
24. Nobili, S., et al.: 16 channels Velodyne versus planar LiDARs based perception system for large scale 2D-SLAM. In: Proceedings of the 7th Workshop on Planning, Perception and Navigation for Intelligent Vehicles (PPNIV). IEEE/RSJ International Conference on Intelligent Robots and Systems, pp. 131–136 (2016)
25. Kudriashov, A., Buratowski, T., Giergiel, M.: Robot's pose estimation in environment exploration process with SLAM and laser techniques. Naukovi Notatki **58**, 204–212 (2017)
26. Moore, T., Stouch, D.: A generalized extended Kalman filter implementation for the robot operating system. In: Proceedings of the 13th International Conference on Intelligent Autonomous Systems (IAS-13). Springer (2014)

27. Meeussen, W.: The robot pose EKF package for ROS source code repository (2018). https://github.com/ros-planning/navigation/. Accessed 29 May 2018
28. Buratowski, T., et al.: Robot z laserowym czujnikiem odległości do budowy map 2D. Modelowanie inżynierskie **30**(61), 27–33 (2016)
29. Grisetti, G., Stachniss, C., Burgard, W.: Improved techniques for grid mapping with Rao-Blackwellized particle filters. IEEE Trans. Robot. **23**(1), 34–46 (2007)
30. Hornung, A., et al.: OctoMap: an efficient probabilistic 3D mapping framework based on octrees. Auton. Robot. **34**(3), 189–206 (2013)
31. Simmons, R., et al.: Coordination for multi-robot exploration and mapping. In: Proceedings of the AAAI Conference on Artificial Intelligence. AAAI, pp. 852–858 (2000)
32. Burgard, W., et al.: Collaborative multi-robot exploration. In: Proceedings of the IEEE International Conference on Robotics and Automation (ICRA), vol. 1, pp. 476–481. IEEE (2000)
33. Özkucur, N.E., Akın, H.L.: Cooperative multi-robot map merging using fast-SLAM. In: Baltes, J., et al. (eds.) Robot Soccer World Cup XIII, pp. 449–460. Springer, Berlin (2009)
34. Thrun, S., Burgard, W., Fox, D.: A real-time algorithm for mobile robot mapping with applications to multi-robot and 3D mapping. In: Proceedings of the IEEE International Conference on Robotics and Automation (ICRA), vol. 1, pp. 321–328. IEEE (2000)
35. Howard, A.: Multi-robot simultaneous localization and mapping using particle filters. Int. J. Robot. Res. **25**(12), 1243–1256 (2006)
36. Fox, D., et al.: Distributed multirobot exploration and mapping. Proc. IEEE **94**(7), 1325–1339 (2006)
37. Carlone, L., et al.: Rao-Blackwellized particle filters multi robot SLAM with unknown initial correspondences and limited communication. In: Proceedings of the IEEE International Conference on Robotics and Automation (ICRA), pp. 243–249. IEEE (2010)
38. Zitova, B., Flusser, J.: Image registration methods: a survey. Image Vis. Comput. **21**(11), 977–1000 (2003)
39. Szeliski, R.: Computer Vision: Algorithms and Applications. Springer Science & Business Media, Germany (2010)
40. Carpin, S.: Merging maps via Hough transform. In: Proceedings of the IEEE/RSJ International Conference on Intelligent Robots and Systems, pp. 1878–1883. IEEE (2008)
41. Carpin, S.: Fast and accurate map merging for multi-robot systems. Auton. Robot. **25**(3), 305–316 (2008)
42. Hörner, J.: GitHub repository multirobot-map-merge package (2018). https://github.com/hrnr/m-explore/. Accessed 20 June 2018

Chapter 5
Evaluation

Abstract This chapter is divided into two parts—a simulation verification and an experimental evaluation of the developed multirobot SLAM-based inspection system. In the first part, several simulation scenes were created in the V-REP environment. This verification covers all parts of the proposed system for both single and multirobot cases. The experimental part has been performed in the form of an inspection and rescue operation in real rough underground terrain, which can be found in coal mines after accidents. For this purpose, the off-road mobile robot platform and a 3D laser rangefinder scanner have been used.

Keywords Multirobot inspection system · Rough terrain inspection · Inspection robotics · Slam · 3d mapping · Isolated terrain

5.1 Introduction

The proposed system evaluation is divided into two parts—simulation along with development and environmental verification. During the simulation part of the evaluation, implementation bugs that have been discovered early as well as method limitations must be fixed in test loops performed after each development increment. This step is called *early testing*. Early testing is closely connected to the proposed system implementation process and generally might be seen as an integral part of that process. Test cases here must be simple and short enough to reach the main goal (along with programming bugs) and take a little time. The goal is to find and potentially fix the most known and often described issues each subsystem is usually facing in the research community. In most of cases, it is enough to verify the subsystem on the function testing level with virtual (mocked or recorded) measurements as well as the behavior of other subsystems. This is usually done by developed software, sometime with dedicated framework usage (unittest, rosplay, etc.). All robot control algorithms and applied method implementations are developed in accordance with the ROS (Robot Operation System) framework [6] standards with ROS community support and library usage. This means that the SLAM-based control solution, which is developed in this book, can be released as an integrated package that could be run

A. Kudriashov et al., *SLAM Techniques Application for Mobile Robot in Rough Terrain*, Mechanisms and Machine Science 87, https://doi.org/10.1007/978-3-030-48981-6_5

on any hardware platform that ROS supports (x386, x64, ARM etc. [6]) and also be easily modified or tuned for very strong specifics of the target task, and used in a wide spectrum of mobile robot platforms (including Unmanned Ground and Aerial Vehicle). That makes the released software independent from hardware specifics.

When early verification is finished, it is time to connect the separate parts together and deliver them to the following step—*simulation verification*. During this part of the evaluation, a fully working system is placed into the virtual environment that simulates real target conditions. A detailed explanation of the chosen methodology of simulation verification and its results may be found in the current section as a separate part. The different cases and tasks are presented there.

The last part of the evaluation is performed in real environment experiments. The release version of the already assembled and virtually tested system is delivered and applied to wheeled mobile robot platforms with a differential drive and four-wheel control. All sensors and measurements are real and all processes, including mapping and path planning, are performed online in real-time. A detailed description might be found in the *Environmental experiments* section.

5.2 Simulation Verification

5.2.1 Introduction

Since all algorithms are implemented inside of the ROS framework, they could be easily tested by a wide range of simulation software. The comparison analysis of simulators used in academic research applications could be found in [7]. In accordance with this survey, the five most often used are: Gazebo, ARGoS, Open Dynamics Engine (ODE), Bullet, and V-REP (Virtual Robot Experimental Platform). All of them can be used to run the ROS application with a different integration difficulty level and a variable range of simulation tools. ODE is generally dedicated to 3D dynamics, while ARGoS and Bullet deliver more universal simulation tools. These three packages are released as libraries that might be imported by third-party calculation or simulation software like Matlab/Simulink. The other two—Gazebo and V-REP—are represented as free, education purpose, standalone applications with a fully functional graphical user interface. Their performance with ROS mobile robot control applications is analyzed in [5]. Regarding the mentioned analysis and the previous experience, it could be concluded that, even though Gazebo is open source and is integrated more deeply into ROS, the V-REP has a more intuitive, user-friendly interface and its integration with ROS reached very comfortable level during the last few years. Therefore, V-REP might be seen as being used more efficiently for complex verification scene creation and simulations of mobile robotic systems like those developed in this book—a SLAM-based inspection system with 3D mapping and dynamic tracking.

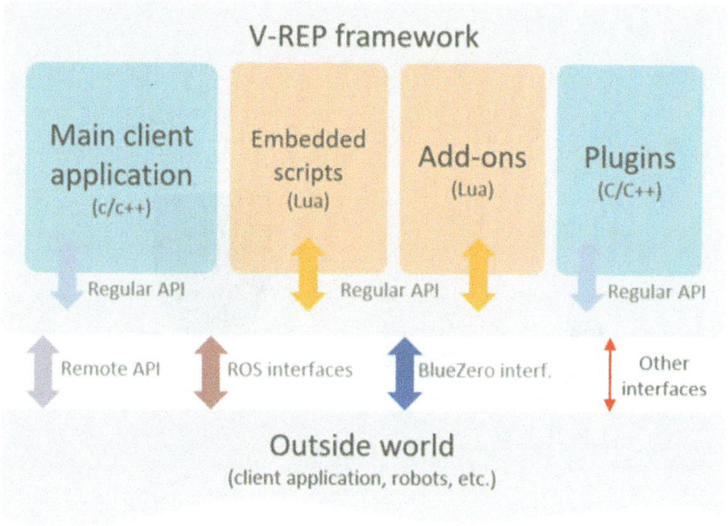

Fig. 5.1 V-REP API framework [1]

5.2.2 V-REP Models and Integration with ROS Applications

The connection between ROS applications and the V-REP simulator environment could be provided by the V-REP API (application programming interface) framework, which includes native and third-party ROS-interfaces and C/C++ API (Fig. 5.1).

For messaging between the control system software implementation parts represented as ROS nodes and simulated mobile robot platform with sensors, the following are used: native RosInterface and RosPlugin along with plugin of Velodyne lidar sensors '**vrep_plugin_velodyne**' implemented in accordance to V-Rep C++ API usage [2] that publishes PointCloud2 messages to ROS topic. Since, in the proposed system architecture, the link between the global and the robot base framework is performed by the use of laser odometry, it is required to have measurements published into the local frame—***'lidar'*** (Fig. 4.11) but the mentioned plugin publishes them into the global frame. This makes it impossible to use that plugin without the required modifications. The fork of the original repository [3] with a minor modification provided fixes this problem, and the Velodyne sensor measurement simulation is performed at a level, which is definitely enough for an efficient scanning simulation. All sensors used during the SLAM-based technique simulation verification are shown in Fig. 5.2. Laser measurements have been delivered to the ROS application by dedicated ROS message frames—LaserScan for 2D and PointCloud2 for 3D scans. IMU is simulated as a set of connected gyroscopes and accelerometers that publishes ROS IMU message. A detailed description of all ROS sensor message frames could be found at the official ROS website [6].

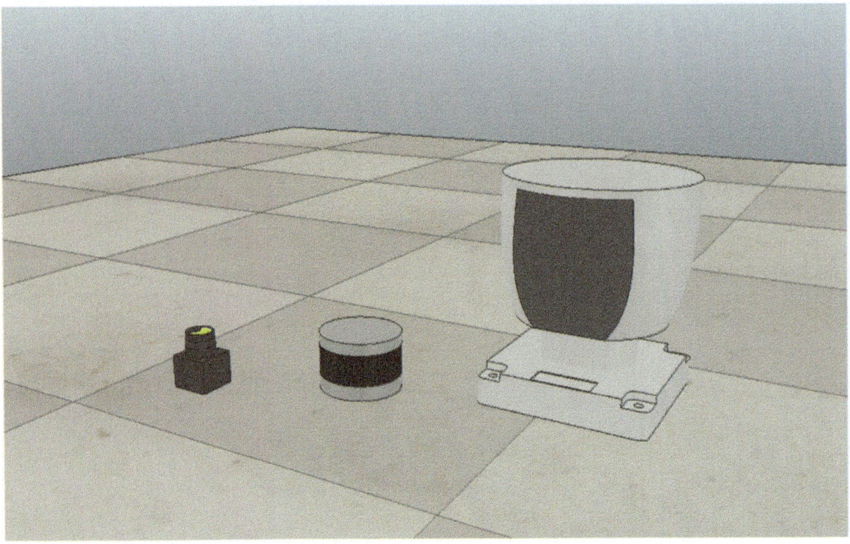

Fig. 5.2 V-REP sensor models used in evaluation: Hokuyo URG 04LX UG01 2D lidar (left), Velodyne VPL-16 (middle) and HDL-64E-S2 (right) 3D lidars

Fig. 5.3 V-REP robotic platform models used in the evaluation: KRiM AGH AntRobot (left) and Pioneer P3DX (right) differential drive controlled mobile robots

Models related to Pioneer P3DX and the differential drive mobile robot developed at the Department of Robotics and Mechatronics at the AGH University of Science and Technology [8, 9] are used as mobile robot platforms (Fig. 5.3).

The control of both robots is provided in accordance to differential drive control fundamentals described in Chap. 2 (Sects. 2.2.2 and 2.4). The world frame linear V_x, V_y and angular θ velocities are given by the applied planner applications or manual control node that publishes velocity messages to ROS '\ ***cmd*** \ ***vel***' topic

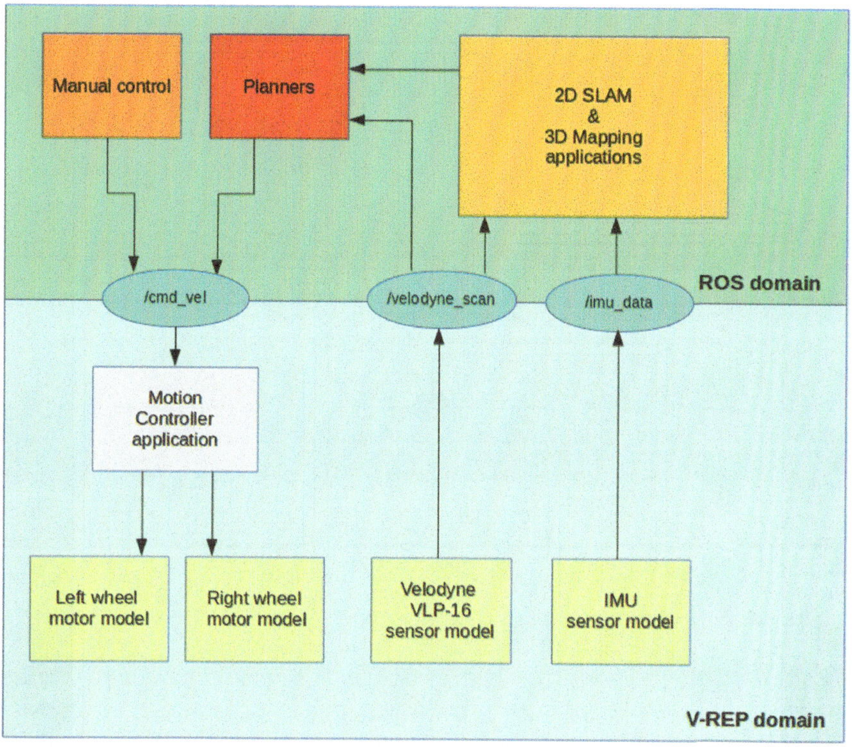

Fig. 5.4 V-REP/ROS mobile robot communication diagram: ROS domain calculates the world frame linear V_x, V_y and angular θ velocities that are sent to '\ ***cmd*** \ ***vel***' topic (top part) and V-REP is listening to this topic along with calculating each wheel velocity in a robot frame (bottom)

(Fig. 5.4). Therefore, the mobile robot controller has to be modified to support control algorithms, which are based on ROS velocity control topic subscription with the solution to differential drive kinematics (2.13) for each wheel velocity calculation as follows:

$$
\begin{aligned}
V_l &= \tfrac{1}{2}(2V_x - L \cdot \theta) \\
V_r &= \tfrac{1}{2}(2V_x + L \cdot \theta)
\end{aligned} \tag{5.1}
$$

where V_l and V_r are linear velocities of the left and the right wheel; L is the distance between these wheels on a common axis.

The software architecture on both sides (ROS and V-REP/mobile robot platform) is developed in accordance with ROS standard; thus, switching between the simulator and the real robotic platform is easy and fast, as the only modification, which may be required, is following the mentioned standards and chosen topic naming conventions (like '\ ***cmd*** \ ***vel***' for controls and '\ ***velodyne_scan***' for measured cloud of points).

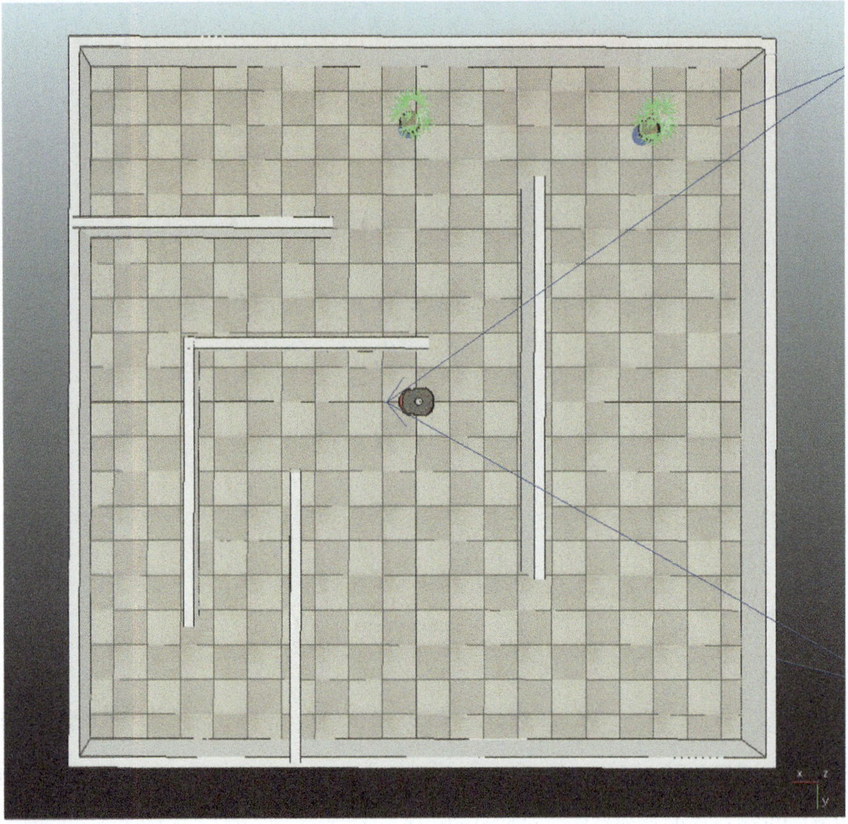

Fig. 5.5 Plane terrain-single agent SLAM simulation evaluation: V-REP scene with a few simple corridors and plants

5.2.3 Plane Terrain Workspace Scene

For the proposed system verification, several scenes that simulate real environment tasks have been created in V-REP. The first and the simplest consists of a few corridors and two plants with a wheeled robot, which is equipped with Velodyne VPL-16 lidar and IMU sensors (Fig. 5.5). An additional simulated sensor is a monocular camera, which is only used for scene monitoring from the ROS RVIZ view when the V-REP sensing data is played by from rosbag file (Fig. 5.6). The use of rosbag allows to repeat the experiment for different SLAM technique applications with the same exact input sensor data that has been generated once by V-REP. In this way, it is possible to compare the method efficiency in the same conditions.

During the simulation verification, the trajectory estimation performance is compared to PL-ICP, RF2O laser odometry and AMCL-EKF filter estimator techniques (Fig. 5.7). As it was assumed, regular pure laser odometry methods are hardly faced

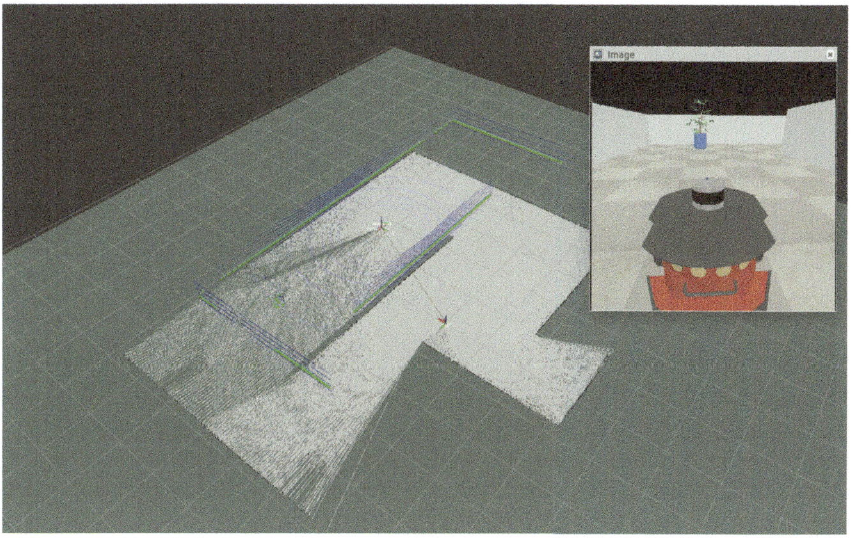

Fig. 5.6 Plane terrain simulation evaluation: RVIZ view during experiments

with an estimation of orientation and motion in long corridors. This issue is reduced a little by combining IMU with PL-ICP laser odometry.

However, the PL-ICP performance generally looks very poor even in a simple plain terrain. This method has huge problems with orientation estimation, which were particularly reduced by IMU, but they still have impact on the whole process. The number of errors is increasing together with the working time (Fig. 5.8) and the average error for each coordinate is larger even by a few times in comparison to the more advanced RF2O technique. The RF2O laser odometry method definitely looks better in the general trajectory comparison, which is reflected by the produced map (Fig. 5.9c,d), but in the orientation estimation, this effect is not as good—the average θ error in this scene is about 0.126 rad, then PL-ICP has 0.4015 rad. The application of the proposed AMCL-EKF filtering technique in the R2FO pose estimation allows to reduce x, y coordinate errors from 0.2998 to 0.066 m for x and from 0.558 to 0.297 m for the y average level, which has a very positive impact on the resulting maps (Fig. 5.9e, f). However, the orientation issue is comparable and the average error of θ in the current scene with a plane terrain may even be bigger by a few times.

The trajectory calculation of absolute errors for all the techniques analyzed in this experiment based on laser odometry can be found in Table 5.1. All presented laser odometry-based techniques perform local pose estimation, as it was already mentioned in Sect. 4.3.1, while presenting the exact pose delivered by V-REP ('world_vrep'). Therefore, it needs to be mentioned again that the trajectory shift between local and global will always be present due to the expected local frame odometry drift. This is why even the ideal local pose filtration will be shifted in comparison to the global reference. This drift is reduced during simultaneous global

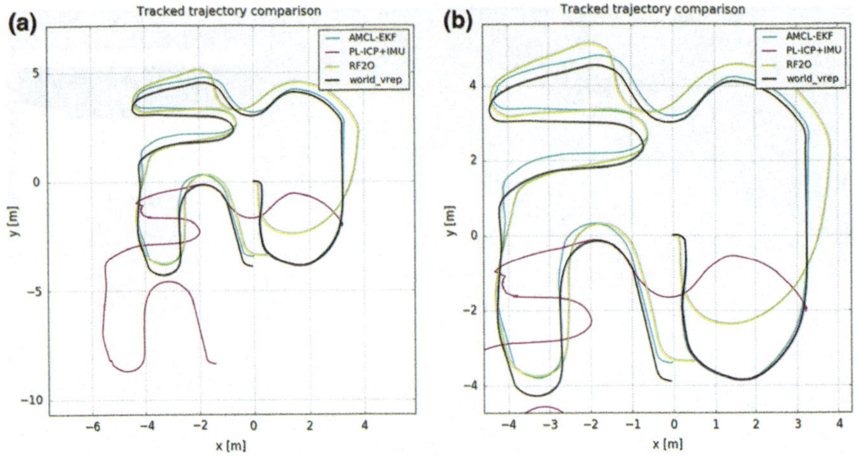

Fig. 5.7 Plane terrain-single agent SLAM simulation evaluation: robot trajectory estimation techniques comparison (**a**) and zoom on AMCL-EKF estimation (**b**)

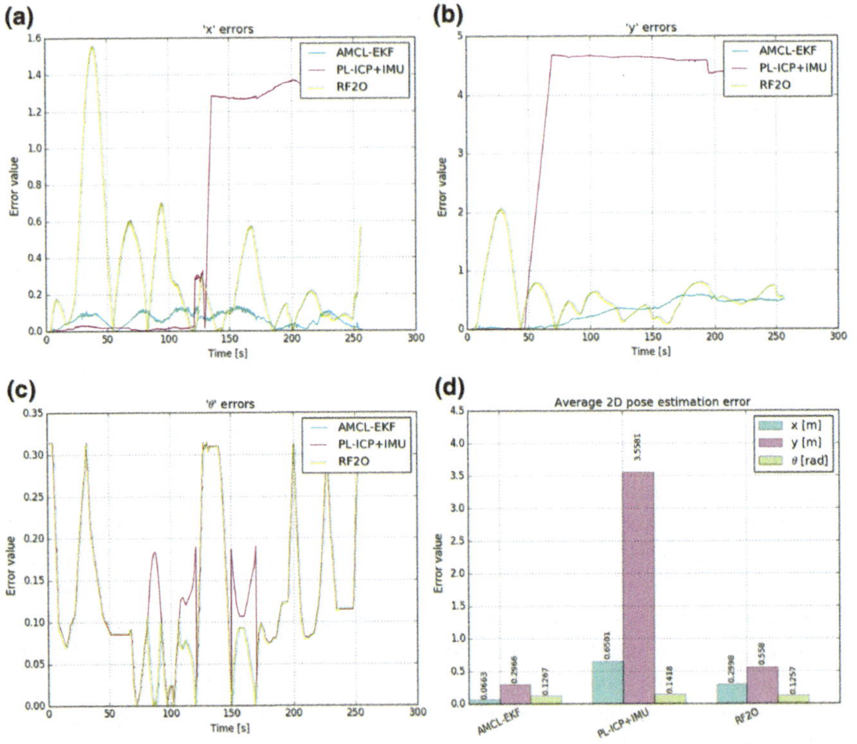

Fig. 5.8 Plane terrain-single agent SLAM simulation evaluation: robot trajectory errors for each coordinates per time—x (**a**), y (**b**), θ (**c**) and average coordinate error (**d**)

Table 5.1 Plane terrain-single agent SLAM simulation evaluation: pose estimation techniques absolute errors comparison

Pose estimation technique	Absolute error								
	x (m)			y (m)			θ (rad)		
	Min	Max	Avg	Min	Max	Avg	Min	Max	Avg
AMCL-EKF	7e-05	0.138	0.066	5e-05	0.599	0.297	1e-06	0.314	0.127
PL-ICP+IMU	1e-05	2.807	0.417	1e-06	6.061	3.34	1.1e-04	0.34	0.14
RF2O	1.5e-04	1.561	0.3	1e-06	2.068	0.558	3e-05	0.315	0.126

localization and map creation, which is performed by Rao-Blackwellized SLAM at '***odom***'—'***map***' connection. Thus, in this way, landmarks measured at the local frame in the resulting map will correspond to real obstacle coordinates on the global frame.

Although, in a simple terrain case, the orientation error is a common issue, which all the presented techniques possess, in the other coordinates, the proposed pose tracking method has much better performance, which is confirmed by map building results (Fig. 5.9).

The maps created in this simulation experiment are shown in Fig. 5.9. It is easy to find that the proposed pose estimation solution has the best performance between in comparison to other performed techniques, which are applied to the proposed combined 2D SLAM/3D mapping system. The 2D map represented as an occupancy grid is more efficient than the regular one projected by the Octomap map framework. In accordance with this Octomap 2D projection, the area of narrow corridors at the bottom left of the map is closed to motion, as it is almost fully occupied by obstacles (Fig. 5.9f), while the 2D SLAM grid map shows it as free spaces (Fig. 5.9e), which is correct. Moreover, since for 2D measurements construction, only obstacles that are discovered inside of precise height level limits are those defined by the robot's $\mathcal{C}$-space, the only obstacles, which can be dangerous for the robot's motion, are presented on the 2D SLAM map. This, in fact, has a positive impact on path planning, as it makes the search for a safe path more robust and efficient. In this way, the assumption that the 2D SLAM map gives more accurate and useful view of workspace information than motion control is confirmed.

The produced 3D map of environments from pre-estimated poses is represented by three-dimensional octrees (Fig. 5.10). This map could be opened and scaled by any software dedicated to octree models. In this research, a free and open-source Octovis software, which was developed by Octomap authors, is used as the octree viewer [4]. During the opened 3D map analysis, it can noticed that all the main features of the 3D scene, including narrow corridors and the mentioned plants, are presented. Moreover, the area, which is marked as occupied in the 2D projection, has a structure relative to the simulated scene and the occupation is presented only at ground level. Thus, it does not disturb the map representation. Therefore, it could

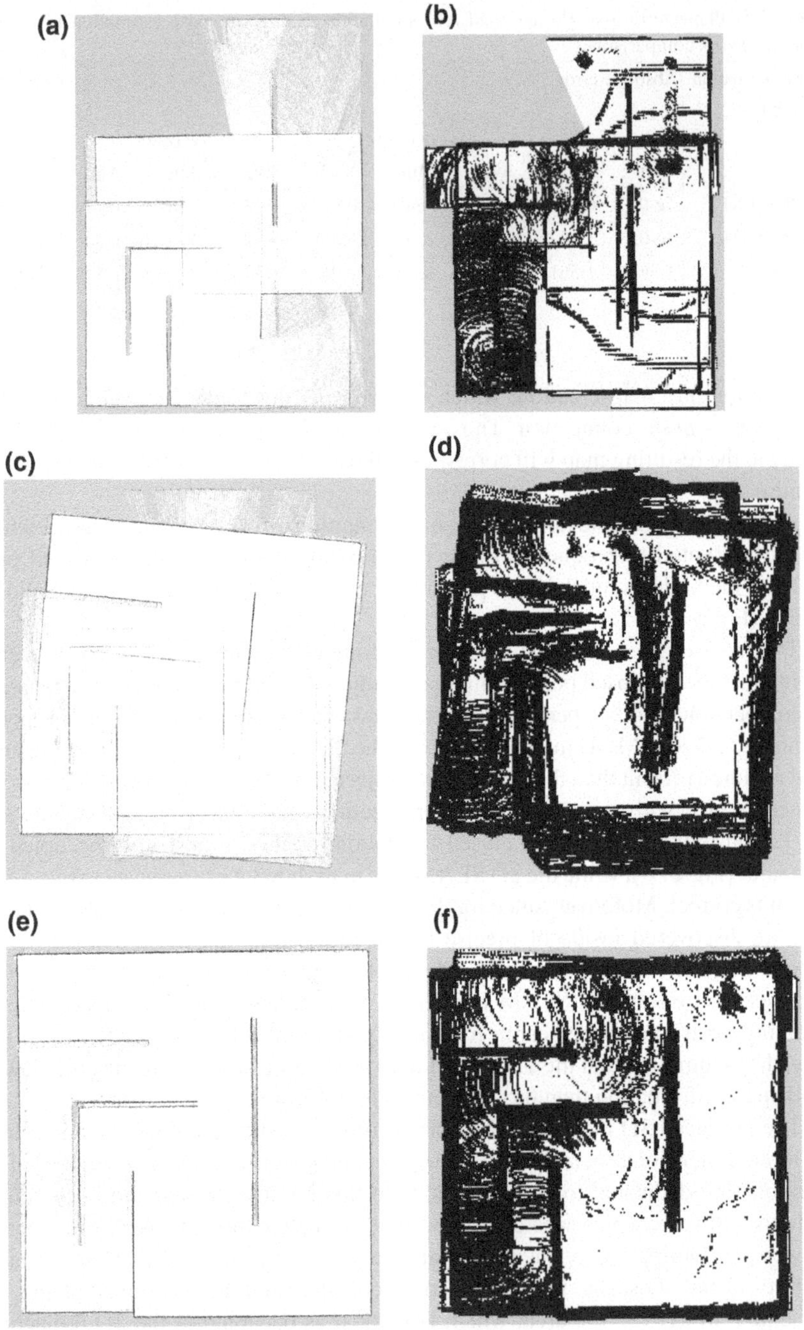

Fig. 5.9 Plane terrain-single agent SLAM simulation evaluation. 2D maps created with different pose estimation techniques: PL-ICP laser odometry with IMU (**a**, **b**), R2FO (**c**, **d**) and the proposed AMCL-EKF technique (**e**, **f**). On left side (**a**, **c**, **e**)—grid maps constructed by 2D SLAM process and on the right (**b**, **d**, **f**)—Octomap projected maps

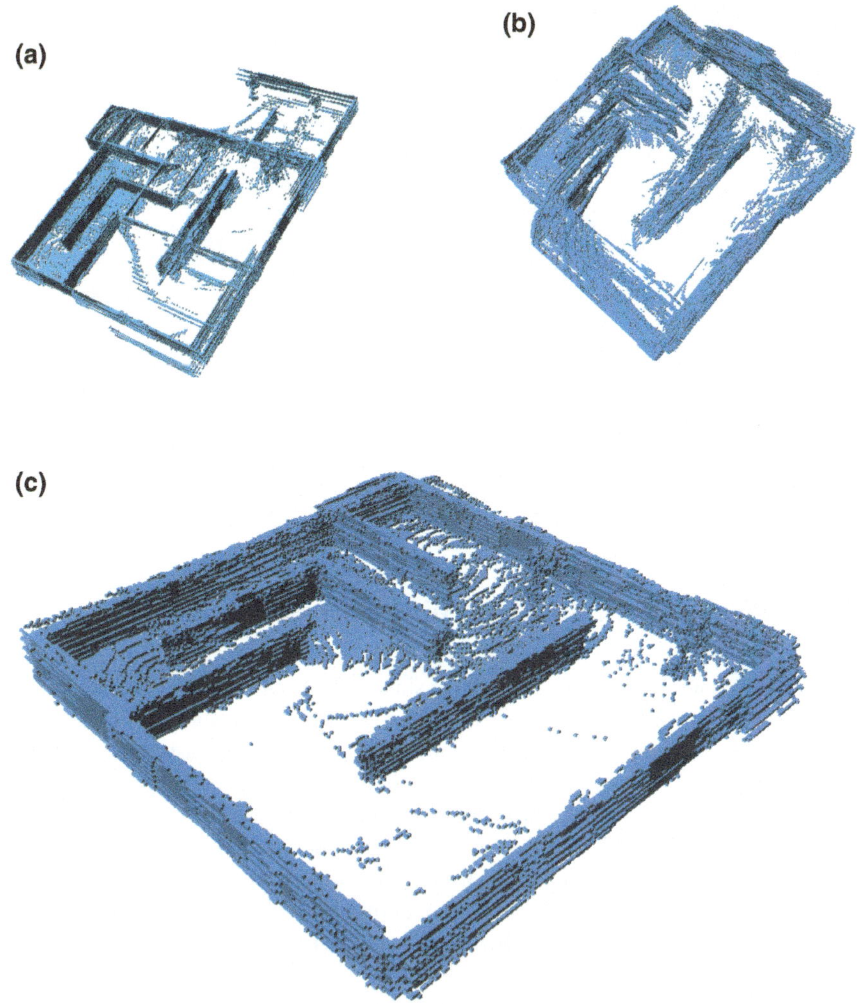

Fig. 5.10 Plane terrain-single agent SLAM simulation evaluation. 3D octree map constructed with the use of PL-ICP (**a**), RF2O (**b**) and the proposed combined 2D/3D AMCL-EKF SLAM technique (**c**) shown by the Octovis software

be concluded that the integrated approach developed in the previous chapter delivers a precise and efficient enough workspace representation that it could guarantee safe and robust mobile robot motion, along with all the main terrain features in 3D maps.

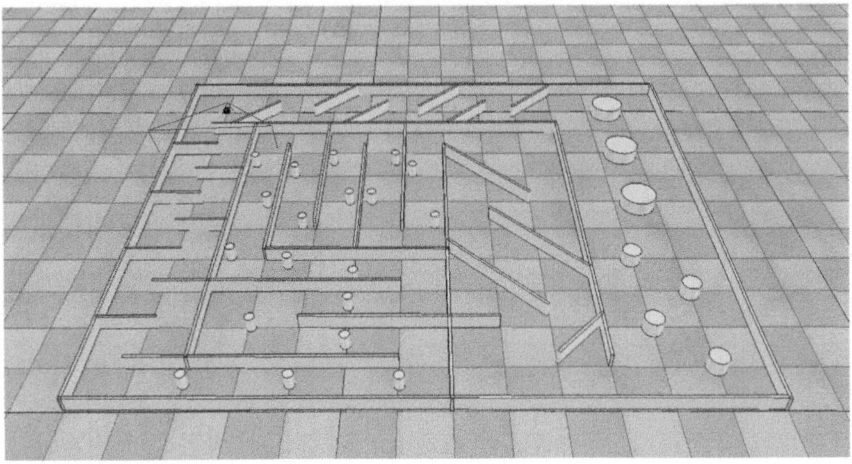

Fig. 5.11 Plane terrain-single agent advanced SLAM simulation evaluation: V-REP scene with large number of corridors and obstacles

5.2.4 *Plane Terrain Advanced Workspace Scene*

The second scene developed in V-REP for simulation verification is dedicated to the proposed system performance check in huge workspaces with a number of different parts that have the same looking features and long distances to be passed. This scene is shown in Fig. 5.11. The same verification method has been applied in this scene as in the case of the previous scene. Since, during this experiment, the biggest errors related to an unsynchronized AMCL-EKF estimator were observed, the accuracy differences were studied in more detailed. For the trajectory comparison of the synchronized and the unsynchronized filter, which was already shown in the previous chapter in Fig. 4.8, the error value calculations could be found in Fig. 5.12 and in Table 5.2.

A comparison of the estimated trajectories by all methods used in this book could be found in Fig. 5.13 and the absolute error calculation in Table 5.3 and Fig. 5.1. As could be seen, the difference between regular RF2O and the proposed AMCL-EKF method in this terrain is not as significant as in the previous case. The average absolute errors for x, y in meters are 0.948 and 2.745 for AMCL-EKF, while for RF2O—1.008 and 2.798. The orientation errors are also comparable—0.102 and 0.103. The pure AMCL technique estimates definitely look the worst in this environment—7.501, 16.146, 0.132 of average errors for all $X^{(2)}$ coordinates x, y, θ, which was caused by an unsynchronized particle filter estimation failure. This could be interpreted in accordance to the whole book narration that finding of a single best solution for all cases is not possible and the selection of methods for an integrated approach should be taken in a specific case. However, even such little differences in pose estimation could have a key impact on map building. In Fig. 5.15, it can be seen that a loop closure problem is present in the 2D SLAM process, which uses RF2O estimates,

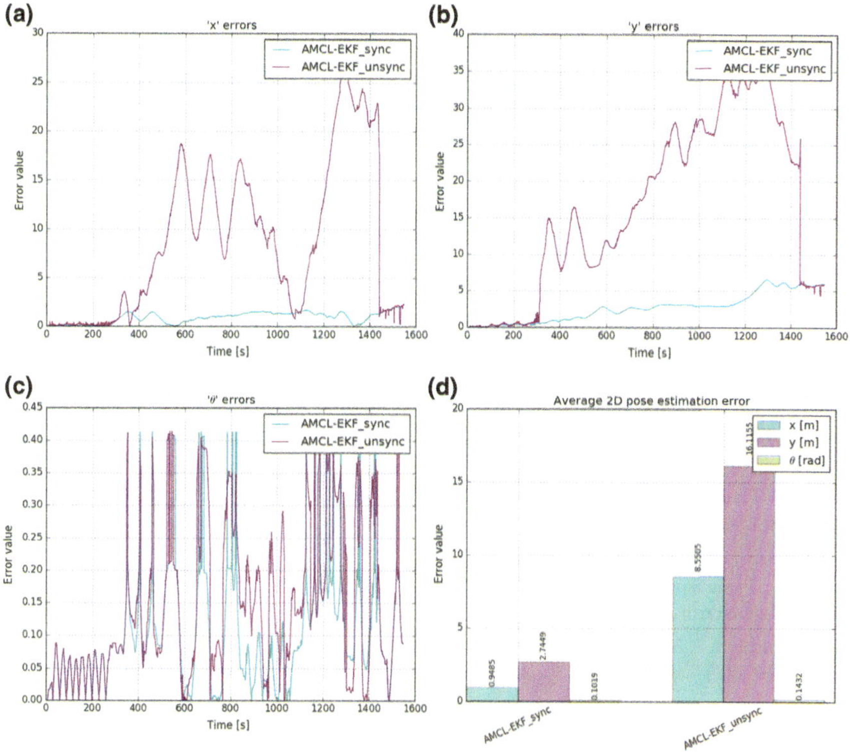

Fig. 5.12 Plane terrain-single agent advanced SLAM simulation evaluation: robot trajectory errors presented in unsynchronized AMCL-EKF (Unsync.) and synchronized AMCL-EKF (Sync.) techniques for each coordinates per time—x (**a**), y (**b**), θ (**c**) and average coordinate error (**d**)

Table 5.2 Plane terrain-single agent advanced SLAM simulation evaluation: unsynchronized AMCL-EKF (Unsync.) and synchronized AMCL-EKF (Sync.) pose estimation techniques absolute errors comparison

Pose estimation technique	Absolute error								
	x (m)			y (m)			θ (rad)		
	Min	Max	Avg	Min	Max	Avg	Min	Max	Avg
Unsync.	9.6e-04	26.75	8.551	2e-05	37.608	16.116	1e-06	0.414	0.143
Sync.	1e-04	2.344	0.941	1e-06	6.556	2.705	1e-06	0.414	0.101

while with pose delivered by the proposed AMCL-EKF technique loop closing is performed pretty well. The 3D map created during the exploration in this workspace is shown in Fig. 5.16. As it was expected, the produced maps could be used as a robust workspace information source (Fig. 5.15).

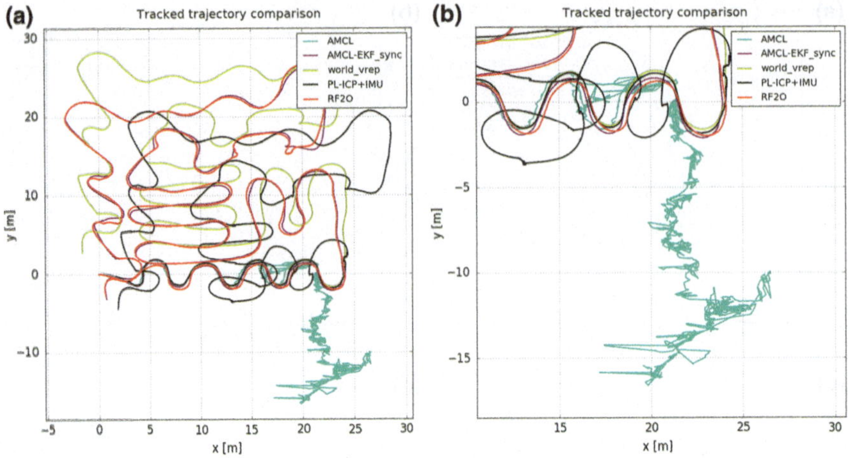

Fig. 5.13 Plane terrain-single agent advanced SLAM simulation evaluation: robot trajectory estimation technique comparison (**a**), crop to AMCL fail (**b**)

Table 5.3 Plane terrain-single agent advanced SLAM simulation evaluation: pose estimation technique absolute error comparison

Pose estimation technique	Absolute error								
	x (m)			y (m)			θ (rad)		
	Min	Max	Avg	Min	Max	Avg	Min	Max	Avg
AMCL	1.7e-04	24.32	7.501	5e-05	42.419	16.146	1e-06	0.413	0.132
AMCL-EKF	4e-05	2.342	0.948	3e-05	6.632	2.745	1e-05	0.414	0.102
PL-ICP+IMU	3e-05	6.843	3.073	3e-05	10.025	6.385	1e-06	0.414	0.127
RF2O	3e-05	2.341	1.008	2e-05	6.767	2.798	1e-06	0.414	0.103

5.2.5 Rough Terrain Workspace Scene

The next experiment is based on motion in rough terrain workspace. The scene for this simulation is similar to the one used in the simple plane terrain evaluation scene (Sect. 5.2.3), but with a rough surface that may cause robot additional dynamic issues and result in complications with both planar and 3D scan data association from the pose tracked in two dimensions (Fig. 5.17). For that type of specific cases, the AMCL-EKF technique based on a laser range-finder and IMU measurements is developed.

According to the comparison of trajectories, which are estimated by the analysis contained in the approaches of this book (Fig. 5.18), it was found that the best performance in this scene with unstructured terrain surface was presented by the developed AMCL-EKF technique, and the second-best performance, as expected—by AMCL. The average absolute errors in meters and radians for these pose estimation techniques

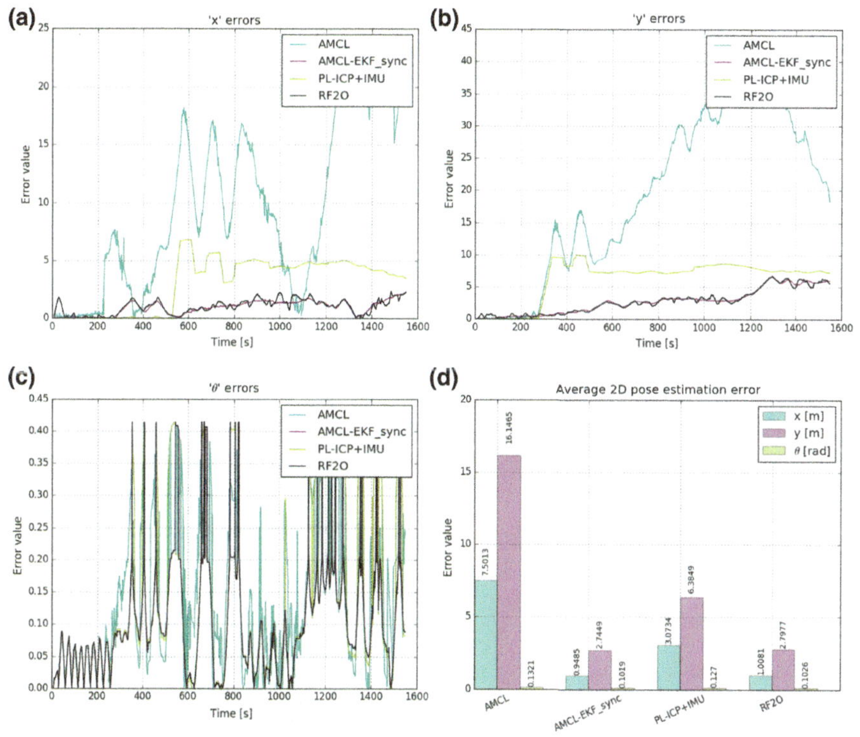

Fig. 5.14 Plane terrain-single agent advanced SLAM simulation evaluation: robot trajectory errors for each coordinates per time—x (**a**), y (**b**), θ (**c**) and average coordinate error (**d**)

with respect to x, y, θ coordinates are 0.316, 0.381, 0.13 and 0.334, 0.404, 0.129, while the next position in the accuracy ranking has 0.494, 0.678, 0.1303 scores and pertains to RF2O (Table 5.4).

So, it could be concluded that, in a rough terrain workspace, the developed AMCL-EKF technique shows the best characteristics that originate from the efficient techniques on which they are based (Fig. 5.19).

The legitimacy of the efficiency of the developed technique application for SLAM-based integrated solutions in an unstructured and rough terrain can be deduced from a map accuracy comparison, which is presented in Fig. 5.20. The constructed 2D map (Fig. 5.20e) is clear and robust, allowing for safe and effective motion, including an autonomous navigation of the inspected mobile robots. Moreover, the workspace representation recreated by the reduced Octomapping algorithm with an application of this exact pose estimation technique is able to represent the three-dimensional characteristics of the workspace more precisely (Fig. 5.21).

The results of this simulation experiment confirm the assumptions and successful implementation of the mobile robot control integration approach developed in Chap. 4, which is based on SLAM-techniques that are selected for work in rough

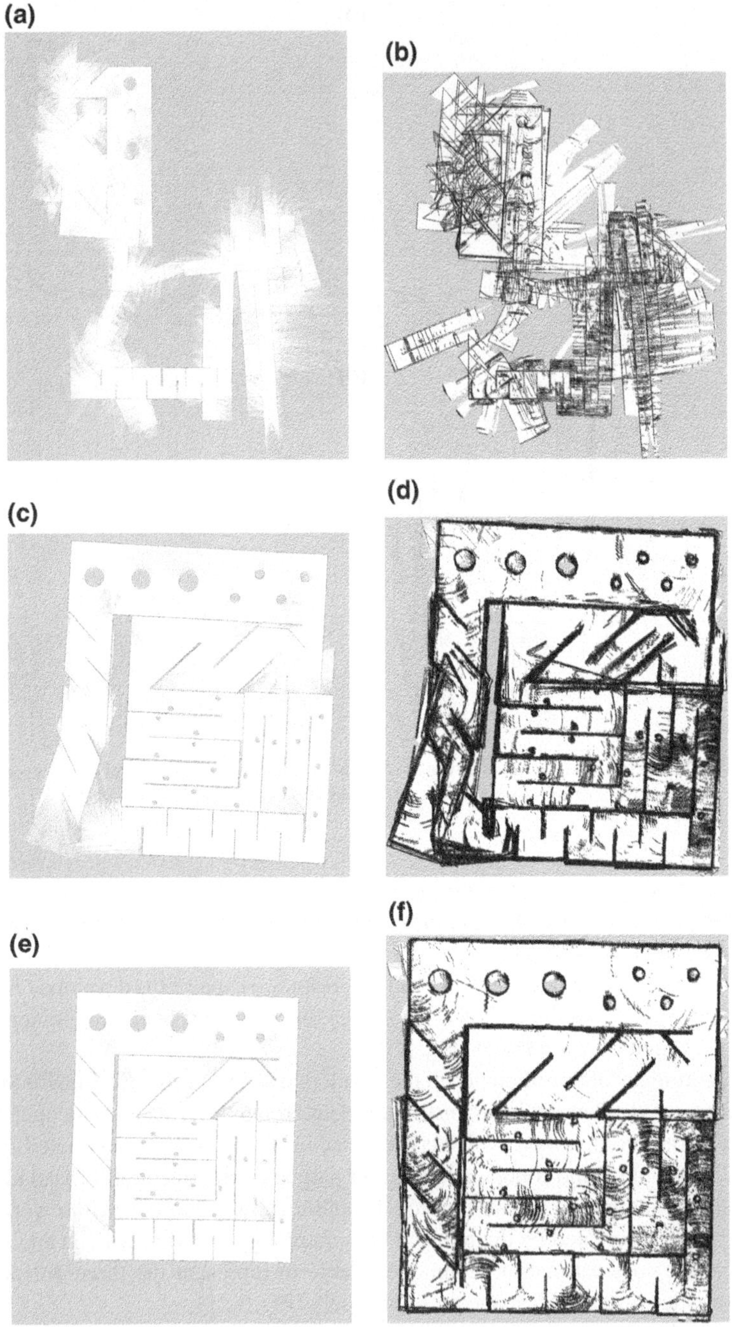

Fig. 5.15 Plane terrain-single agent advanced SLAM simulation evaluation. 2D maps created with different pose estimation techniques: PL-ICP laser odometry with IMU (**a**, **b**), R2FO (**c**, **d**) and the proposed AMCL-EKF technique (**e**, **f**). On the left side (**a**, **c**, **e**)—grid maps constructed by the 2D SLAM process, and on the right (b, d, f)—Octomap projected maps

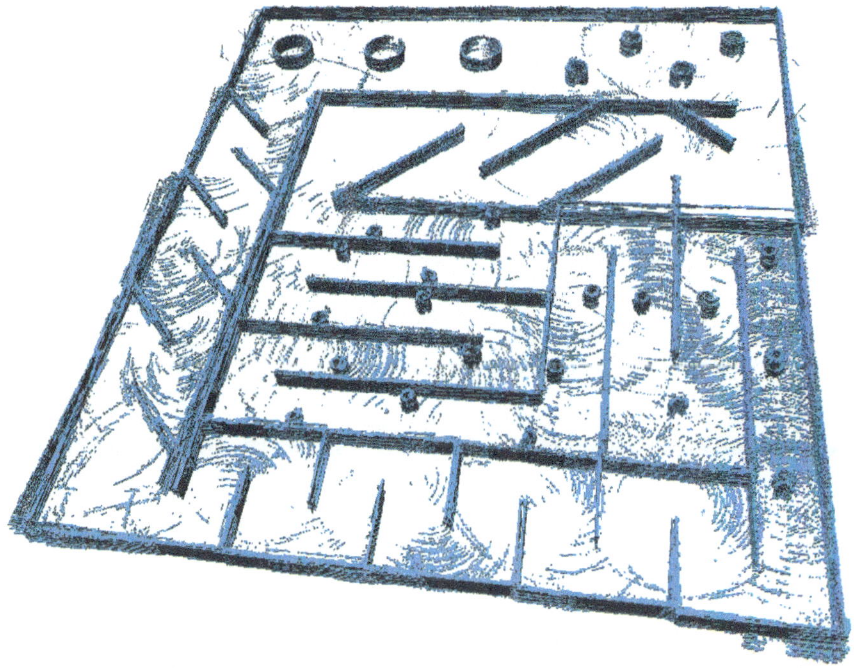

Fig. 5.16 Plane terrain-single agent advanced SLAM simulation evaluation. 3D octree map constructed with the proposed combined 2D/3D SLAM technique shown by the Octovis software

terrain. So, it could be concluded that the developed single robot system is ready for the last stage of the evaluation—the experimental verification in real, hard-to-access and rough surface terrain. This experiment description and its results can be found in the second part of this chapter (Sect. 5.3).

5.2.6 Multirobot Exploration

The last part of the simulation evaluation is dedicated to the verification of the proposed multirobot system that connects the developed integrated SLAM-based mobile robot control system with the group of the homogenous approach to mobile robots. The multirobot approach development with a detailed discussion is described in the previous Chap. 4 (Sect. 4.4). The development was based on a simple V-REP scene with three mobile robots that have the same configuration and are controlled by a hierarchical hybrid path planning approach that consists of the global planner A* algorithm implementation and dynamic obstacle avoidance, which is solved by the local reactive potential field planning technique (Fig. 4.24). Since the robots' cooperation and map fusion results were already shown before (Fig. 4.25), in this

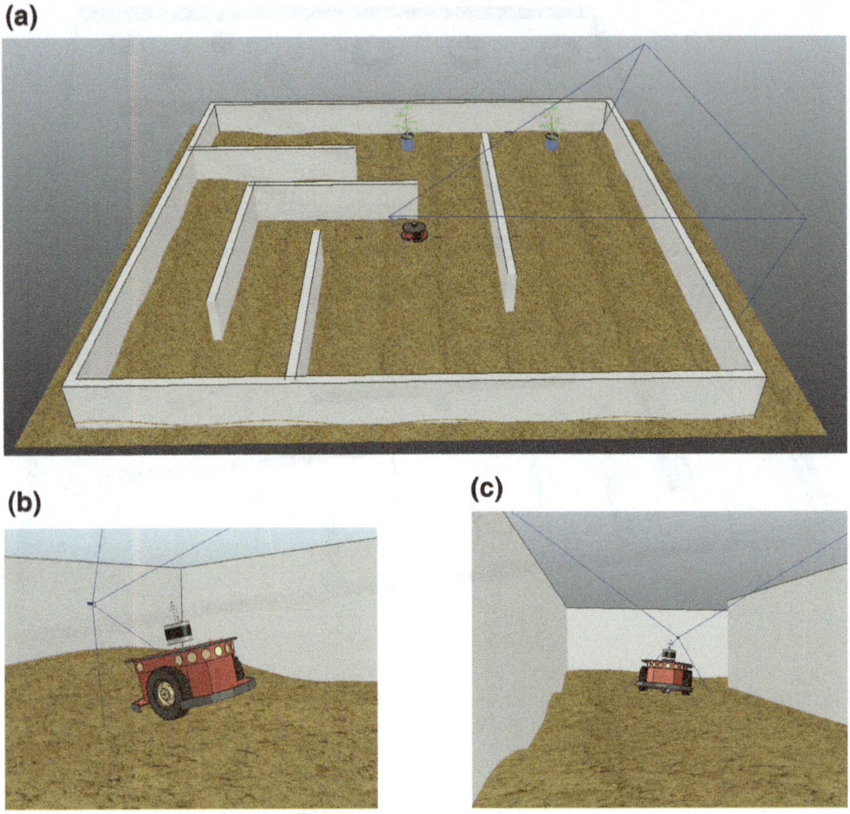

Fig. 5.17 Rough terrain-single agent SLAM simulation evaluation: V-REP scene with rough off-road terrain (**a**), differential drive mobile robot motion in rough terrain (**b**, **c**)

chapter, we only will look at the experiment results from the passed trajectory and the efficiency of the path planning solution.

Let us take a look how could the path planning behavior differ in a very close environment and with the application of the same algorithms. Each robot in a group received a motion target at the opposite side of the workspace, and the planning system was started (Fig. 5.22).

The full exploration time is about 460 s and can be divided into three parts—beginning, middle and ending. At the initial and first part, all robots present a comparable expansion (Fig. 5.22); the observable differences are too small and could be easily missed. But the 'middle' time performance after 190 s passed shows that, while one robot has just exited the narrow passage (Fig. 5.23c), the other has already reached the target position (Fig. 5.23b). This diversity can be interpreted as the same approach used in robotic systems with identical configurations and a similar common environment, where the results would be the same. Therefore, it is very hard to find, which path planning approach is the best solution. The technique selection should

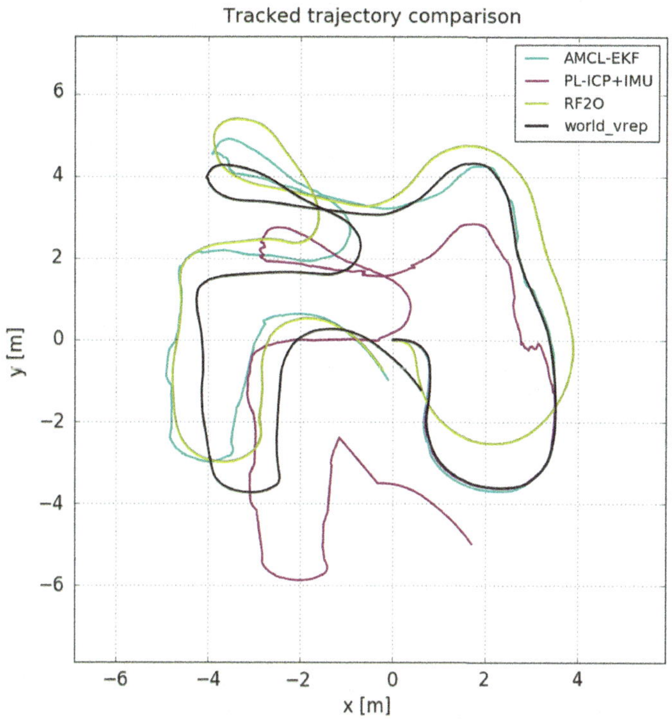

Fig. 5.18 Rough terrain-single agent SLAM simulation evaluation: robot trajectory estimation technique comparison

Table 5.4 Rough terrain-single agent SLAM simulation evaluation: pose estimation techniques absolute error comparison

Pose estimation technique	Absolute error								
	x (m)			y (m)			θ (rad)		
	Min	Max	Avg	Min	Max	Avg	Min	Max	Avg
AMCL-EKF	1e-05	1.072	0.316	1e-06	1.104	0.381	3.9e-04	0.321	0.13
PL-ICP+IMU	3e-05	1.22	0.644	1e-06	3.762	1.458	7e-05	0.315	0.14
AMCL	1e-05	1.876	0.334	6e-05	2.195	0.404	1.1e-04	0.375	0.129
RF2O	2e-05	1.309	0.494	1e-06	1.884	0.678	4e-04	0.322	0.1303

be taken on the basis of a broad and deep study of all the integrated approach system parts and the dedicated terrain specification with a choice of a navigation approach that would be good enough to guarantee a safe motion from the current position to the target.

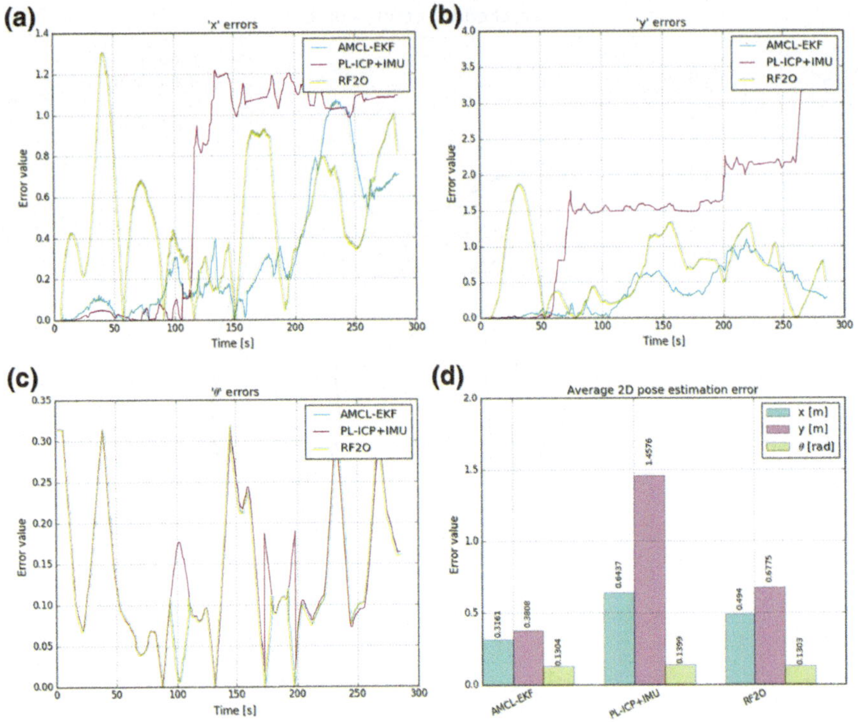

Fig. 5.19 Rough terrain-single agent SLAM simulation evaluation: robot trajectory errors for each coordinates per time—x (top left), y (top right), θ (bottom left) and average coordinate error (bottom right)

All robots finally met at about 380s, where the 'rendezvous' event described in Chap. 4 (Sect. 4.4) has been performed (Fig. 5.24). The result of this event is common with the one shown in Fig. 4.25 from the previous chapter. During meetings, other robots are considered as dynamic obstacles that are added into temporary local cost maps (Fig. 5.25), which allows to avoid a potential collision and have the path planning approach decentralized without any reservation and prioritizing mechanisms (Fig. 5.24).

The full trajectory passed by all robots can be found in Fig. 5.26.

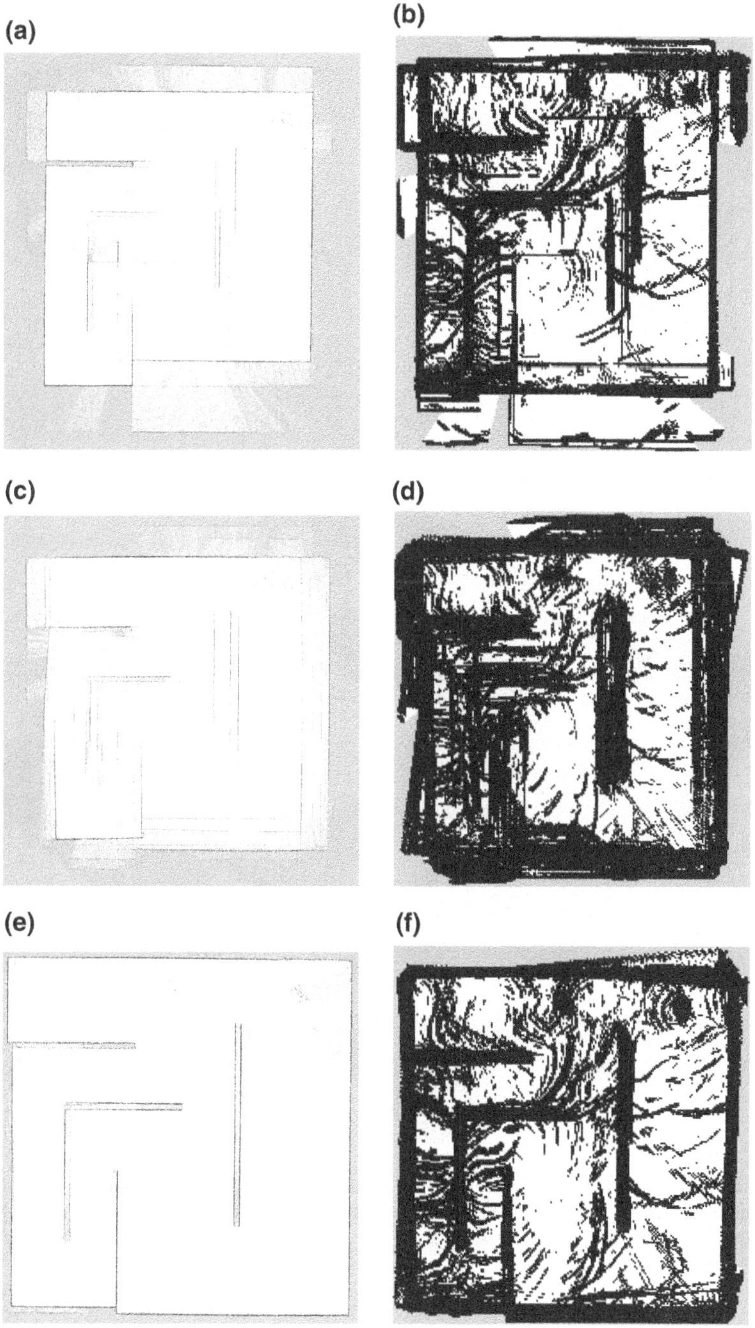

Fig. 5.20 Rough terrain-single agent SLAM simulation evaluation. 2D maps created with different pose estimation techniques: PL-ICP laser odometry with IMU (**a**, **b**), R2FO (**c**, **d**) and the proposed AMCL-EKF technique (**e**, **f**). On the left side (**a**, **c**, **e**)—grid maps constructed by the 2D SLAM process, and on the right (**b**, **d**, **f**)—Octomap projected maps

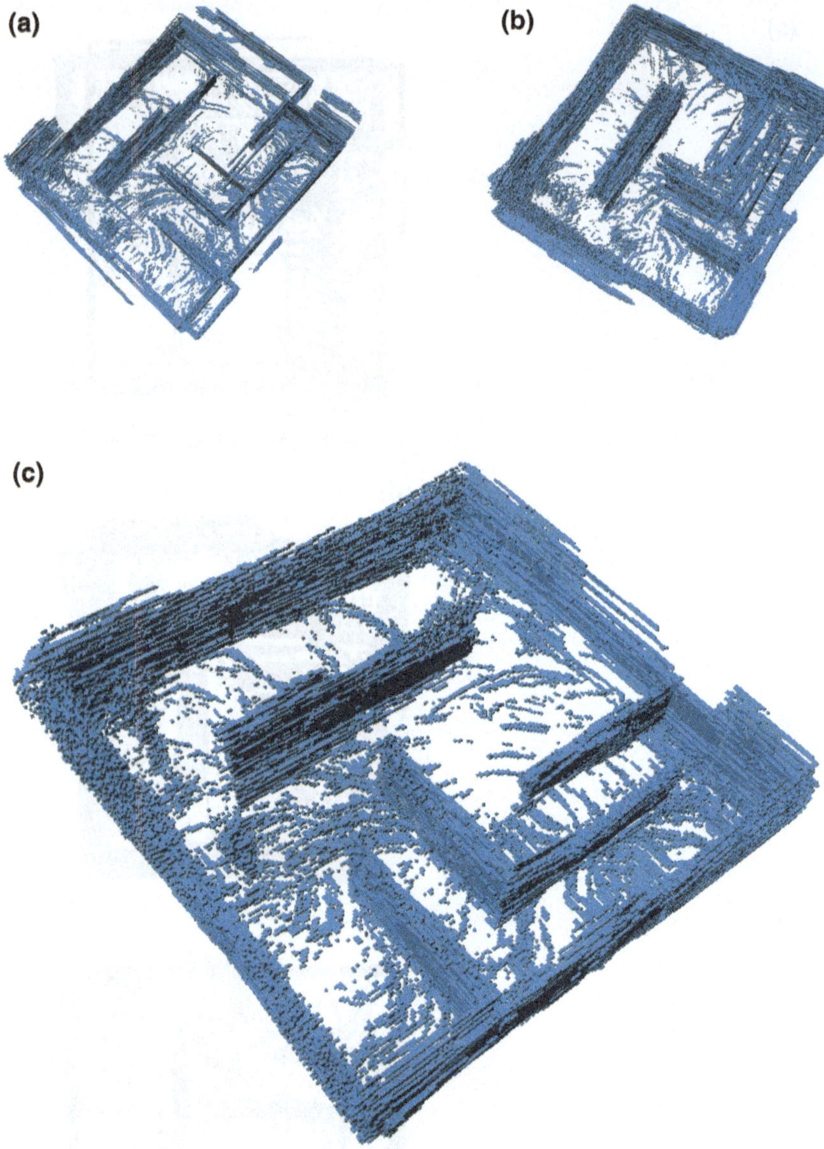

Fig. 5.21 Rough terrain-single agent SLAM simulation evaluation. 3D octree map constructed with the use of PL-ICP (**a**), RF2O (**b**) and the proposed combined 2D/3D AMCL-EKF SLAM technique (**c**) shown by the Octovis software

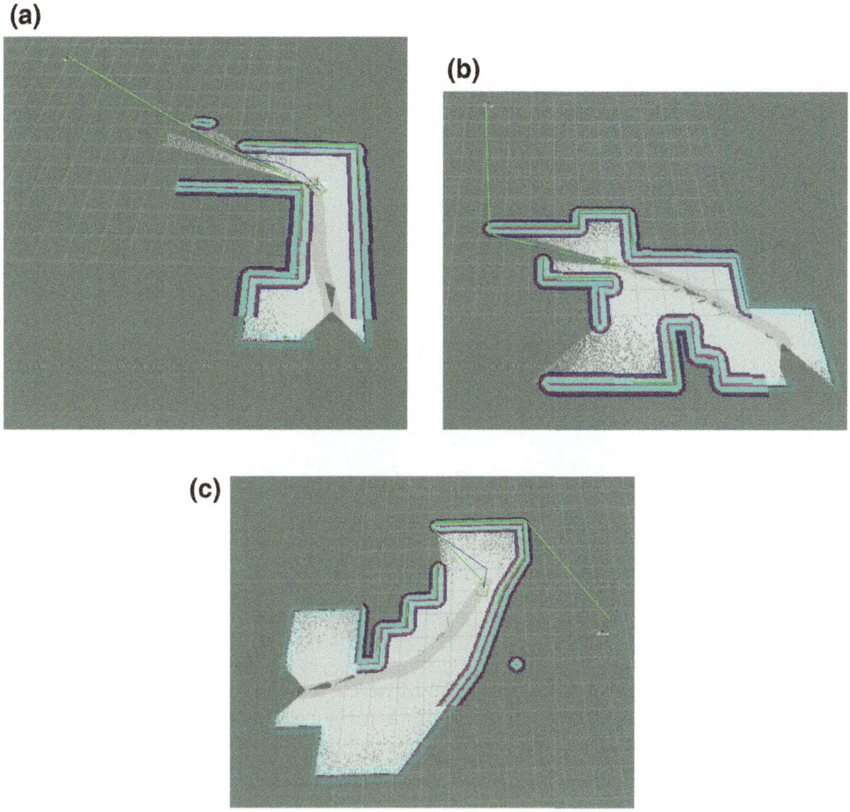

Fig. 5.22 Multirobot exploration evaluation: RVIZ view at robot no. 0 (**a**), 1 (**b**) and **c** state after 50s of unknown space simulation. The global plan is performed by the A* algorithm (green) and the local one by the potential field planning (blue)

5.3 Environmental Experiments

5.3.1 Introduction

The current section is a final part of the developed system approach evaluation, which is based on a performance verification during a real-world task. For this purpose, two mobile robot platforms were involved: the Dr Robot Jaguar 4×4 outdoor mobile robot developed by the Canadian company Dr Robot equipped with a Velodyne VPL-16 3D LIDAR sensor (Fig. 5.27a) and a differential drive mobile robot developed at the Department of Robotics and Mechatronics at the AGH University of Science and Technology group of mobile robots [9] with a Hokuyo UTM-30LX-EW planar LIDAR (Fig. 5.27b).

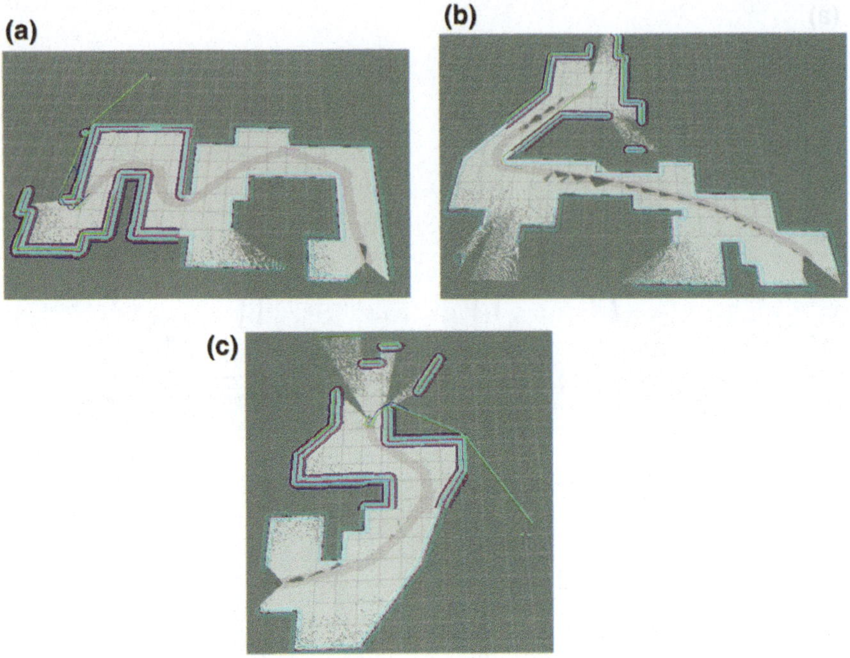

Fig. 5.23 Multirobot exploration evaluation: RVIZ view at robot no. 0 (**a**), 1 (**b**) and **c** state after 190s of unknown space simulation. The global plan is performed by the A* algorithm (green) and the local one by the potential field planning (blue)

Since the quality evaluation was already performed in a previous section—the simulation verification—this part is only focusing on a practical application of the developed integrated approach system. Therefore, the presented case is the underground inspection in a potentially dangerous and hard-to-accesses terrain that includes rough surfaces (Sect. 5.3.2). The experiment was performed by cooperation with the Department of Mining, Dressing and Transport Machines at the AGH University of Science and Technology in the experimental mine, which is situated on the AGH University campus. The description of this inspection and its results can be found below.

5.3.2 Underground Inspection in Mine-Like Rough Terrain

This experiment is an implementation of the inspection and rescue operation scenario with a single robot in an underground mine terrain after an accident, where human presence is very dangerous. The operation assumes motion on both smooth and rough surfaces without the possibility of using any other sensors apart from a laser scanner and IMU.

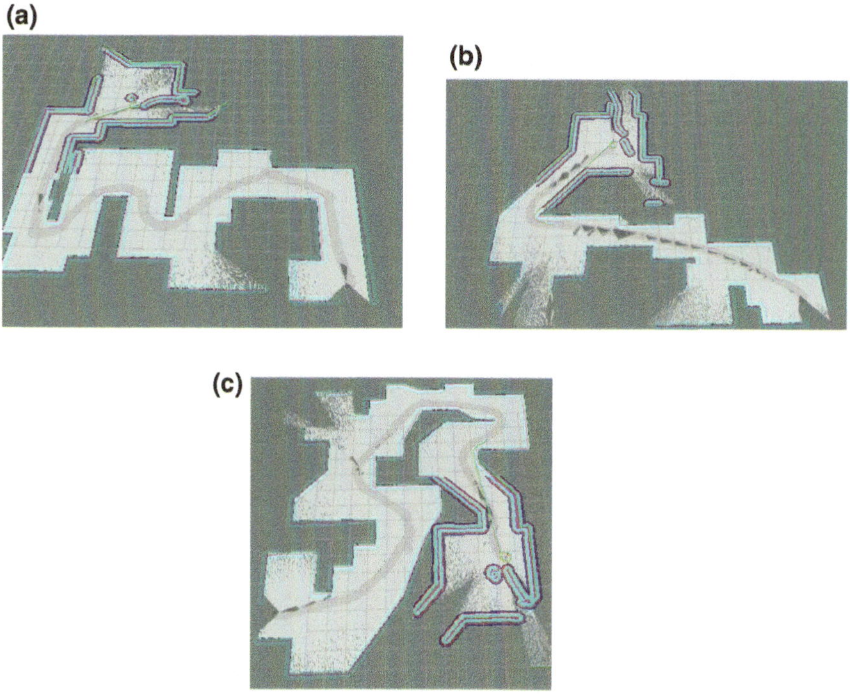

Fig. 5.24 Multirobot exploration evaluation: RVIZ view at robot no. 0 (**a**), 1 (**b**) and **c** state after 380s of unknown space simulation. The global plan is performed by the A* algorithm (green) and the local one by the potential field planning (blue)

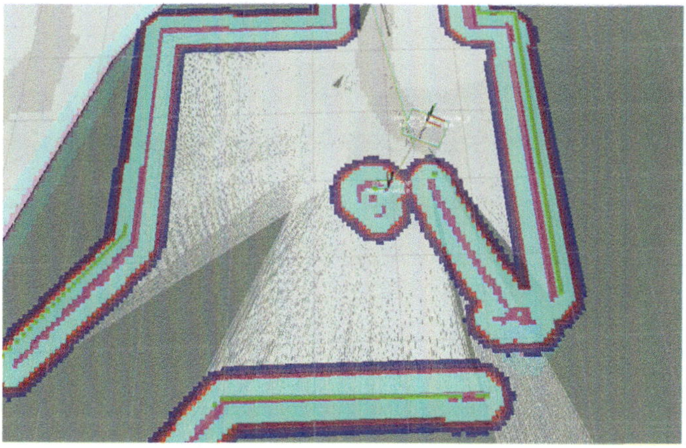

Fig. 5.25 Multirobot exploration evaluation: Considering other robots as obstacles on a local costmap in the path planning system of one of the robots

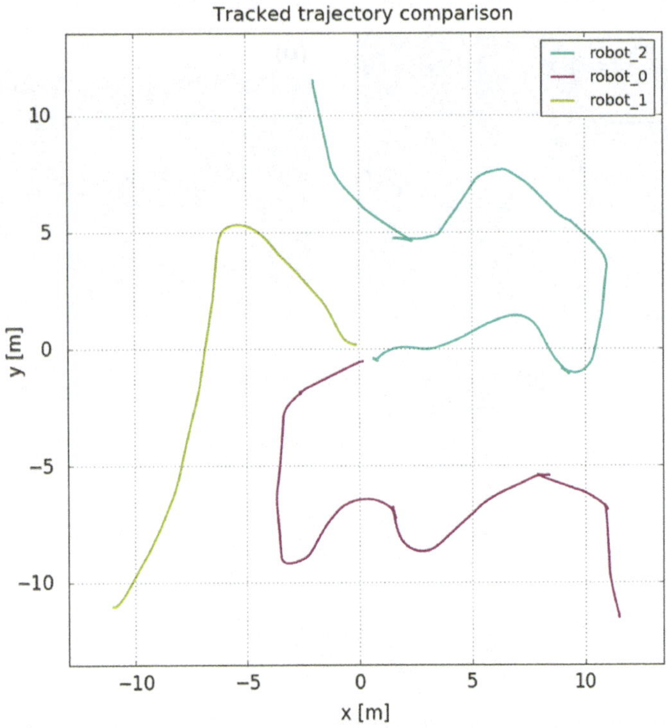

Fig. 5.26 Multirobot exploration evaluation: Trajectories passed by three homogeneous robots

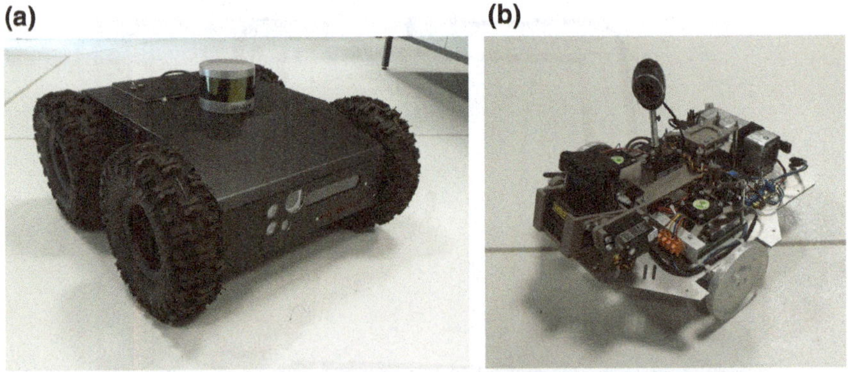

Fig. 5.27 Mobile robotic platforms used in the experimental verification: DrRobot Jaguar 4 × 4 (**a**) and differential drive mobile robot developed in the Department of Robotics and Mechatronics AGH University of Science and Technology (**b**)

Fig. 5.28 Underground inspection in mine-like rough terrain

In Fig. 5.28, several fragments of the operational world have been snapshotted. As we can see, some parts of the road are easy for motion (Fig. 5.28a, f), other give a good example of rough terrain (Fig. 5.28c–e), while in specific places where the accident has occurred, the road was completely closed-off (Fig. 5.28b, c).

Regarding the robot's construction specifics, the inspection was divided into two separate parts. A low level and a high level. This separation caused that, in one place, the presented 'stairs' were too high for the robot's wheels to be able to move through. Therefore, unfortunately, a manual delivery of the robot's base to the railway track (Fig. 5.28d, e) was necessary.

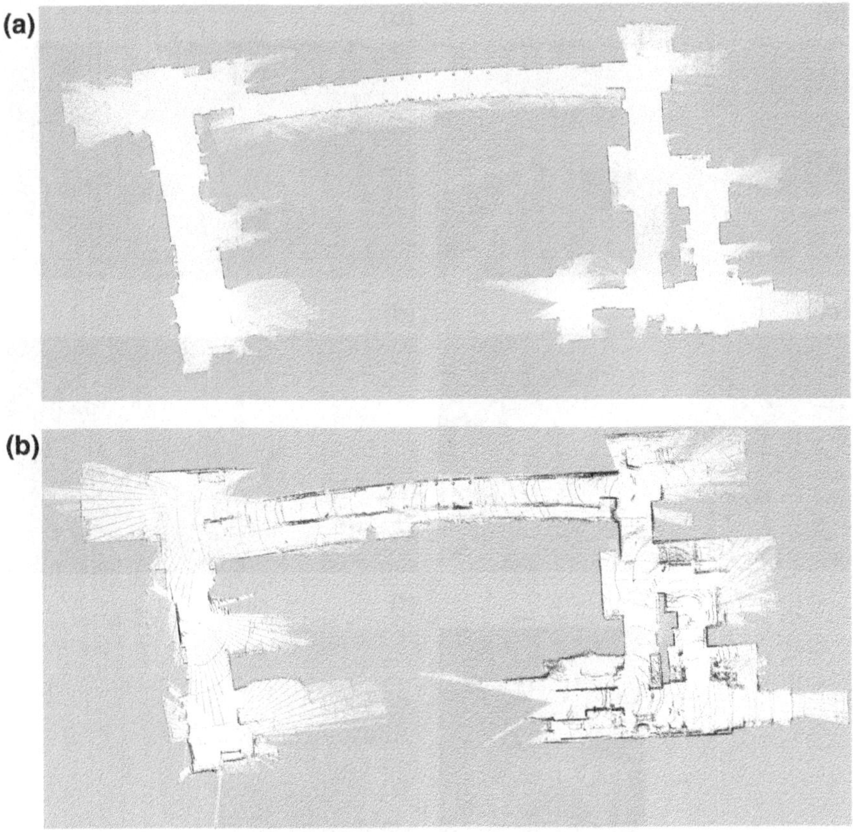

Fig. 5.29 Underground inspection in mine-like rough terrain. 2D maps created with proposed AMCL-EKF technique (**a, b**). On left side (**a**)—grid maps constructed by 2D SLAM process and on right (**b**)—Octomap projected maps

The results of the first part are shown in Fig. 5.29. Both 2D and 3D maps precisely recreates the real state of mine corridors with all observable main features. The skipped part is under rocks (Fig. 5.28b) with the rest of the space occupied by the above-mentioned railway track. The 2D maps estimated by the proposed technique and constructed by the octomapping process differ in the conclusion of occupancy. While the Octomap is showing the actual picture of the workspace actual (Fig. 5.29), the occupancy grid map estimated by the Rao-Blackwellized SLAM approach is focused on only free spaces that are used for further safe navigation. The octomap is used for a precise environment inspection in three dimensions and includes all features of the corridors that have been passed (Fig. 5.30).

While the terrain presented thus far in this experiment is mostly related to smooth-structured surfaces (with a few exceptions), the motion on the railway tracks must be definitely considered as a hard-to-access, rough surface workspace. The resulting

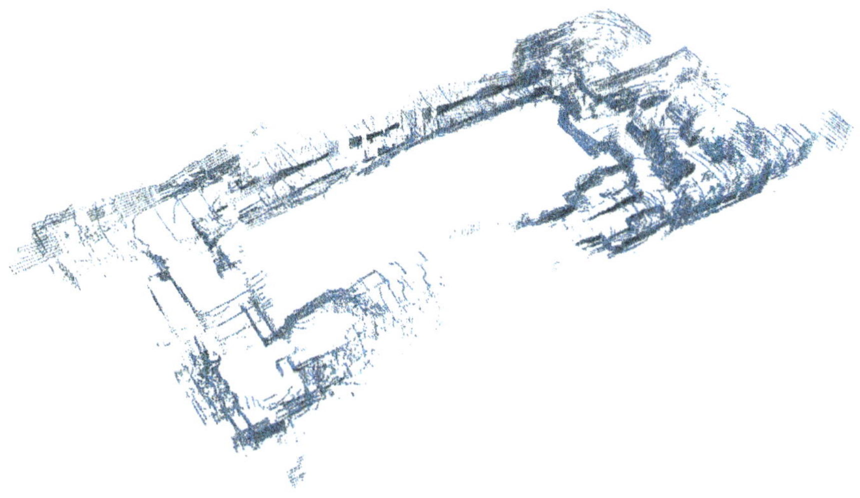

Fig. 5.30 Underground inspection in mine-like rough terrain. 3D octree map constructed with proposed combined 2D/3D SLAM technique shown by Octovis software

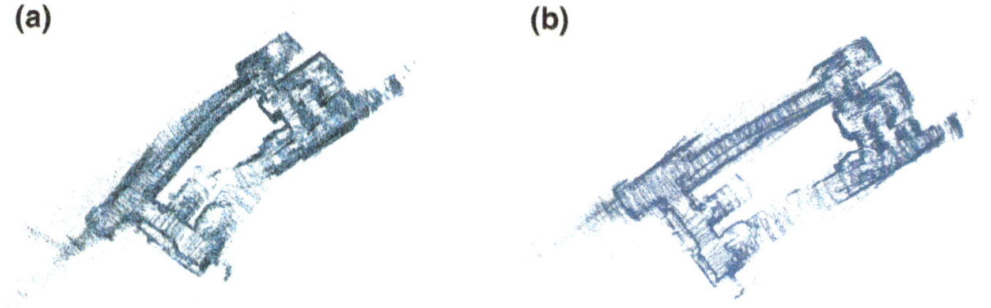

Fig. 5.31 Underground inspection in mine-like rough terrain. 3D octree maps of the main part constructed with the PL-ICP (a) and RF2O (b) pose estimation techniques shown by the Octovis software

2D maps are shown in Figs. 5.33 and 5.34. Here, the differences in the focus of maps can be clearly seen. A big part of the terrain is excluded from the estimation made by the 2D SLAM technique, as there was no possibility to perform the motion, while the octomap projection shows that it is less accurate with an unclear bolder line dividing it, which can cause different confusions in map interpretation. Moreover, the application of the second map to the path finding solution can be very dangerous and the calculated path might include poses that may injure or even destroy the mobile platform. Thus, the usage of the proposed map technique for the navigation approach is strongly recommended.

The 3D terrain map is precise enough to find where the accident has been happened, and how the actual state of the environment looks. The connection of the 2D and the 3D workspace representation gives broader and more useful information on

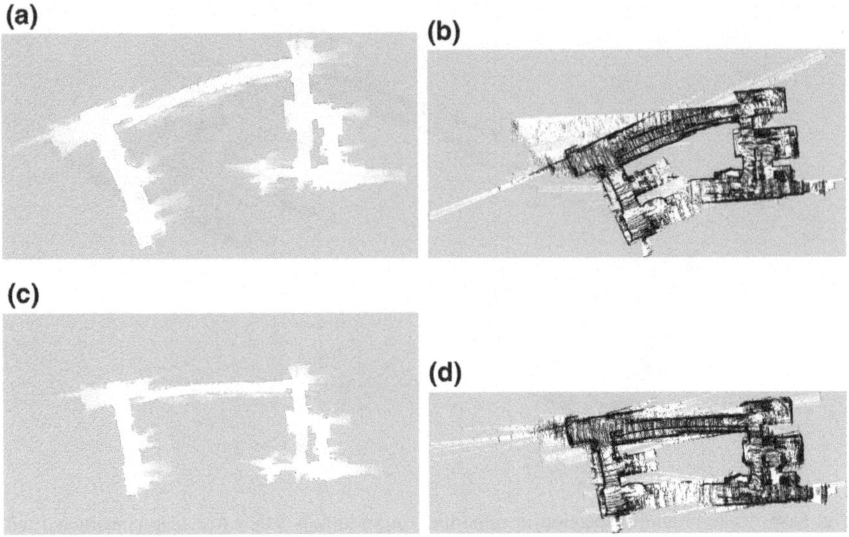

Fig. 5.32 Underground inspection in mine-like rough terrain. 2D maps of the main part, created with the PL-ICP (**a**, **b**) and RF2O techniques (**c**, **d**). On the left side (**a**, **c**)—grid maps constructed by 2D SLAM process, and on the right (**b**, **d**)—Octomap projected maps

the explored environment, so the best and highly recommended solution is to have both map creation techniques deployed on the mobile robot inspection platform.

The experiment presented thus far was prepared online using both 2D and 3D map creation with full performance verification.

During the inspection, the measurement data represented as a time stamped cloud of points was collected, allowing to prepare the map creation comparison, similar to the one shown in the simulation verification part. The 2D maps created with the use of the RF2O and PL-ICP laser odometry techniques could be found in Fig. 5.31, and the 3D octomaps for the mentioned cases are shown in Fig. 5.32. The comparison of other techniques to the ones developed in this book research once again confirms the increase of accuracy and practical usage legitimacy of the proposed systems and their advantages over others.

5.4 Summary

The evaluation presented in this chapter confirms the efficiency, accuracy and robustness of the developed integrated SLAM-based approach for mobile robots control in rough terrain.

The results presented in simulation part show that the developed AMCL-EKF local pose or odometry estimation technique has the best performance in a rough

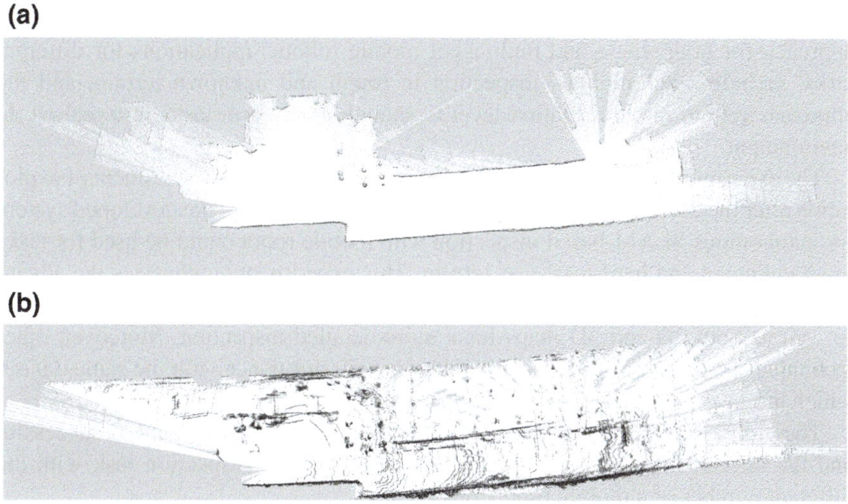

Fig. 5.33 Underground inspection in mine-like rough terrain. 2D maps of the difficult-to-access part, created with the proposed AMCL-EKF technique (**a**, **b**). On the left side (**a**)—grid maps constructed by 2D SLAM process, and on the right (**b**)—Octomap projected maps

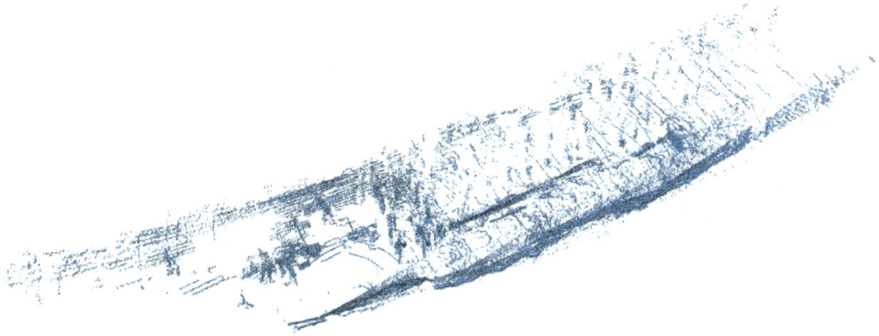

Fig. 5.34 Underground inspection in mine-like rough terrain. 3D octree map of the difficult-to-access part constructed with the proposed combined 2D/3D SLAM technique shown by the Octovis software

unstructured terrain, while in the structured terrain its estimate errors are comparable to RF2O technique. Moreover, the usage of unsynchronized filter might cause estimation disaster. This issue was already described and solved in Chap. 4.

A separate part of the simulation evaluation covers the discussion on the efficiency of the proposed multirobot inspection approach which is homogeneous in behavior and based on hierarchical path planning. All simulations have been performed in V-REP environment, which has been chosen based on simulation tools used in robotics research works survey.

All reached results confirm the legitimacy of the use of the developed integrated approach for both single and multiagent mobile robotic applications for different tasks, including autonomous inspection in rough and unknown terrain, and the approach system implementation level is enough to perform tasks in a real-world environment.

The experimental verification is represented by the prepared experiment of exploration after the accident in the underground mine. It shows that the developed system for autonomous SLAM-based inspection with mobile robot could be used for tasks in an unknown and hard-to-access terrain. This experiment emphasizes the advantages of the map which is produced by 2D SLAM with 3D map usage for navigation as well as both 2D and 3D maps for a more detailed inspection. Moreover, other techniques have demonstrated a certainly worse performance with the same dataset which has been collected during the experiment.

Therefore, the results gathered in this chapter could be interpreted as successful and the developed system can be recommended for the inspection task with the different number of a mobile robots involved.

References

1. Coppelia robotics website: V-REP API framework (2018). http://www.coppeliarobotics.com/helpFiles/en/apisOverview.htm. Accessed 31 May 2018
2. GitHub repository: 'vrep_plugin_velodyne' ROS plugin software repository (2018). https://github.com/omcandido/vrep_plugin_velodyne. Accessed 31 May 2018
3. GitHub repository: 'vrep_plugin_velodyne' ROS plugin software repository - local frame (2018). https://github.com/tigerk0430/vrep_plugin_velodyne. Accessed 31 May 2018
4. Hornung, A. et al.: OctoMap: an efficient probabilistic 3D mapping framework based on octrees. Auton. Robots **34**(3), 189–206 (2013)
5. Nogueira, L.: Comparative analysis between gazebo and v-rep robotic simulators. In: Seminario Interno de Cognicao Artificial-SICA 2014, p. 5 (2014). Chapter 5. Evaluation 123
6. ROS website: About ROS (2018). http://www.ros.org/about-ros/. Accessed 31 May 2018
7. Ivaldi, S., Padois, V., Nori, F.: Tools for dynamics simulation of robots: a survey based on user feedback (2014). arXiv:1402.7050
8. Buratowski, T., Uhl, T., Chmaj, G.: The Project of autonomous group of 2-wheeled mobile robots. In: Proceedings of the 12th World Congress in Mechanism and Machine Science, Besançon, France (2007)
9. Buratowski, T., Ciszewski, M., Giergiel, M., Kudriashov, A., Mitka, Ł.: Robot z laserowym czujnikiem odległości do budowy map 2D. Modelowanie inÿżynierskie **30**(61), 27–33 (2016)

Chapter 6
Conclusions and Future Work Discussion

Abstract This chapter closes the book research by summarizing the results and discussing other conclusions. A possible future work direction and research plans can be found in chapter.

Keywords Multirobot inspection system · Slam · Mapping · Mobile robotics · Navigation · Rescue robots

6.1 Conclusions Discussion

The SLAM-based mobile robot control system presented in this book, which is represented by an integrated approach to autonomous inspection tasks, has been successfully developed and verified. One of the main focuses of the performed research was on the pose estimation approach and its impact on the efficiency of the applied SLAM techniques. The proposed technique connects two separate estimation paradigms—Kalman and particle filtering—into the novel local odometry estimation approach. The approach is based on LIDAR and IMU sensing, along with local Monte-Carlo localization and the laser odometry technique for local pose estimation by an Extended Kalman Filter. This allowed to merge skid and slippage resistance presented in AMCL, precise dynamic orientation changes tracking in three dimensions, which are delivered by IMU with the drive type independence of laser odometry method by multidomain measurements estimation in EKF for an efficient pose tracking. The tracked pose is defined on the yx plane. The three-dimensional coordinates required for the 3D mapping reconstruction are performed by the proposed mechanism from local Eulers angles and z coordinate that are merged with an estimated global pose. In rough terrain, this approach looks robust and definitely more accurate in comparison to other analyzed techniques, while in workspaces with plane-structured terrain, the performance differs—in some cases, it is much better, and in others, it is comparable.

A. Kudriashov et al., *SLAM Techniques Application for Mobile Robot in Rough Terrain*, Mechanisms and Machine Science 87, https://doi.org/10.1007/978-3-030-48981-6_6

The next part of the research was focused on mapping techniques. Along with regular 2D and 3D maps, the original 'planar 3D' world representation was introduced. This concept allowed to reach precise 2D SLAM performance in a specific terrain with multiple levels, like multi-floor buildings. Since this and the regular 3D mapping requires pose tracking by six coordinates (relative to the axes and Euler angles), the 3D mapping was solved by the continuous estimation technique based on an occupancy grid with the octrees representation. The 3D map technique is a reduced Octomap algorithm with the exact pose delivered by the 2D SLAM and 3D pose recreation. This map is used for workspace representation with a lot of information needed for further analyses on the inspection and state monitoring perspective. Safe and efficient motion is guaranteed by a more convenient and accurate planar map, calculated by the 2D SLAM process based on Rao–Blackwellization. The advantages and legitimacy of this approach were confirmed by both simulation and underground mine experiments.

A separate part is applying the developed approaches into a multirobot inspection task system. The system proposition and its implementation consider the cooperation between robots in a group from the homogeneous point of view; thus, the integration of the developed SLAM-based system was simple in implementation, since it assumes that all robots have the same configuration and behavior. The interaction between robots is performed during sporadic meetings—rendezvous events. The algorithm of common map creation was implemented and successfully verified. The map fusion is based on considering occupancy grid maps as images, so the image processing techniques, with Hugh transformation matrix for the calculation of a common coordinate frame, were applied. The merging process is based on data exchange that assumes knowledge of the recent pose and the explored map of each robot involved in the rendezvous.

Multirobot motion planning is solved by the use of the hybrid hierarchical approach. Applying the hierarchical approach allows to ignore complicated and very difficult in implementation multirobot collision avoidance techniques family, which assume a decoupled or centralized path, or part of the path reservation with a strong and defined priority policy. These kind of 'reservation' solutions have a huge number of requirements for the robot configuration and the target workspace, which makes the application of such a complicated solution in rough terrain almost impossible. Therefore, the proposed approach development is based on the application of two planners—the global plan is calculated through the implementation of the A* algorithm that calculates the shortest possible path from a given workspace knowledge, while safe motion is delivered by local potential field planning, which follows the global plan by incremental movement—the goals are taken from the path calculated by A*. Since the potential field planning is a reactive approach, the other robots are considered as dynamic obstacles and they are avoided. The weakest point of this technique is the possibility of a robot being stuck in one place for a long period of time, which is common for potential field algorithms for motion through narrow passages. This means that the efficiency of robots expansion or the time required to reach the target position can be different for the same robots from a group in a similar environment. This was observed and discussed during the evaluation part.

The evaluation includes both a simulation and a real task implementation experiment. After a survey on the most known and often used research simulation environment in mobile robotics, the V-REP simulator has been chosen. The simulation verification was performed on several scenes that include both structured and unstructured terrain. After this, the developed system was applied to an off-road mobile robot platform—DrRobot Jaguar 4 × 4—and used for an inspection and rescue operation simulation in rough underground terrain of an experimental mine located at the Faculty of Mechanical Engineering and Robotics of the AGH University of Science and Technology in Krakow. The experiment results confirm and fulfill the robustness of the developed SLAM-based integrated mobile robot approach for inspection tasks in a rough terrain. Estimated by the proposed method, the 2D and 3D maps give precise information on the state of the mine corridors after an accident. While the 2D SLAM map is more accurate and focused on guaranteeing safe motion, the 3D octomap and its 2D projection provide more helpful information on current state of the terrain. However, applying only the 2D projection from the octomap as a world representation for mobile robot navigation may be risky because of the imprecision of the created projection, which has unclear borders that can cause confusion about the map interpretation. The application of a path planner in these places can be very dangerous because the calculated path might include poses that may injure or even destroy the mobile platform. Therefore, the best practice is to have both maps created.

The integrated approach presented in this book—a mobile robot inspection system along with its multirobot extension—could be applied in several cases for different tasks. First, and the most obvious in a multirobot inspection system, is a rapid expansion of unknown terrain. Applying a number of robotic explorers significantly speeds up world discovery in comparison to a single robot exploration. A group of mobile robots could be deployed in a target workspace with expansion algorithm applications with a similar basis as the one described in Chap. 2 RRT [1], and consequently, discover the full possible C_{free} space. This task assumes that the target terrain is unknown, and the map exchange and merge technique, which is described in the previous section, will unquestionably decrease exploration time. Next, a possible application for the proposed system could be found in rescue or search operations in a terrain difficult to access or potentially dangerous for human operations. This case assumes both known and unknown terrain maps and could be related to war conflict zones or city terrains after an accident. An injured person may send rescue requests to all robots available in range and they will rapidly search for a shortest, safe path to the target, together with a terrain inspection that might help in the risk assessment.

Based on all of the above mentioned, it can be concluded that the SLAM-based system for mobile robot control in rough terrain development performed in this book is a working solution, which can be adapted to different tasks performance, with the main focus on motion and inspection in rough and hard-to-access terrain.

6.2 Future Work Discussion

Based on the results and challenges during the SLAM-based mobile robot inspection system development, the two main fields of future research can be defined—an improvement of combined 2D/3D SLAM technique, and the development focused on a multiagent inspection approach.

The first field research can cover the optimization of global pose estimation technique as well as the precision of 2D mapping technique. The global pose estimation is currently performed by adapted 2D SLAM technique, which is based on Grisetti's algorithm. Therefore, the additional research under the global pose, described by six coordinated, shows that the estimation may increase the accuracy of localization and mapping.

The discussed in this book 2D map was mostly considered from the safe and efficient motion planning perspective, while a fuller image of the workspace in two dimensions was delivered by planar projection from the 3D map. This projection includes a large number of errors and its interpretation may be difficult in many cases. The 2D map from proposed SLAM technique is more accurate but skips the terrain that cannot be used for safe motion. As it was mentioned in book conclusions, the best way is to use both of them for a more efficient inspection. So, it could be valuable to find the map construction technique which might allow to have more information from both inspection and navigation perspective in one map. Therefore, it could be one of the possible future directions.

Another, mapping improvement could be contributed to 3D world representation by a set of planar maps approach, which has been developed and presented in this book. Here, the technique of the loop closure between two planar maps of each level could be developed. Currently, they are connected by only continuous $\mathbf{X}^{(3)}_{glob}$ pose tracking, the connection by global locations will allows to increase path planning efficiency in this approach.

Multirobot approach presented in this book defines robot's cooperation and behavior as homogeneous, therefore, separate directions may cause the development of the decoupled heterogeneous solution which is based on the proposed technique for a single robot.

Since map exchange and merge has been solved only for 2D occupancy grid maps, the other development may be taken on an octree map fusion and common map of the target environment recreation in three dimensions.

The planning approach proposed in this book is hierarchical and based on A* and potential field planning. As potential field planner, as many other approaches, is facing with motion problems in narrow passages, the local planning technique could be improved by the dedicated planner development.

Finally, it is very important to mention that from the research developed in this book, we may conclude that none of the proposed future work direction can be applied in all of the possible cases, and any improvement and technique development must be considered from the target application perspective.

Reference

1. Tang, X., Lien, J.-M., Amato, N.M.: An obstacle-based rapidly exploring random tree. In: 2006 Proceedings of the IEEE International Conference on Robotics and Automation (ICRA), pp. 895–900. IEEE (2006)

The manufacturer's authorised representative in the EU is Springer Nature Customer Service Centre GmbH, Europaplatz 3, 69115 Heidelberg, Germany. If you have any concerns regarding our products, please contact ProductSafety@springernature.com

Printed and bound by CPI Group (UK) Ltd, Croydon, CR0 4YY
30/09/2026
02225665-0004